T0298641

Advanced Mammalian Cell Culture Techniques

This up-to-date book compiles both basic and advanced laboratory techniques of mammalian cell culture. It is divided into four major sections encompassing the basics of cell culture, nucleic acid and protein isolation, cell-staining techniques, and cell transfection and single-cell analysis. The topics include aseptic handling, media preparation, and passaging of cells. The book also outlines downstream assays such as nucleic acid and protein isolation from *in-vitro* cell cultures.

Key Features:

- Covers cellular staining using fluorescent dyes, genetic manipulation of cells via transfection, and an introduction to single-cell analyses
- Discusses basics in cell culture and downstream applications including gene and protein expression analysis
- Includes the principles underlying each of the techniques and provides a detailed methodology to practice
- Explores the whole range of techniques – from basic to downstream applications and advanced methods

The book is essential for students and researchers in the field of life sciences, biotechnology, genetics, and molecular biology.

Advanced Mammalian Cell Culture Techniques

Principles and Practices

Edited by
K.M. Ramkumar
R. Senthilkumar
Md Enamul Hoque

CRC Press
Taylor & Francis Group
Boca Raton London New York

CRC Press is an imprint of the
Taylor & Francis Group, an **informa** business

Designed cover image: K.M. Ramkumar, R. Senthilkumar, and Md Enamul Hoque

First edition published 2024
by CRC Press
2385 NW Executive Center Drive, Suite 320, Boca Raton FL 33431

and by CRC Press
4 Park Square, Milton Park, Abingdon, Oxon, OX14 4RN

CRC Press is an imprint of Taylor & Francis Group, LLC

© 2024 selection and editorial matter, K.M. Ramkumar, R. Senthilkumar, and Md Enamul Hoque; individual chapters, the contributors

Reasonable efforts have been made to publish reliable data and information, but the author and publisher cannot assume responsibility for the validity of all materials or the consequences of their use. The authors and publishers have attempted to trace the copyright holders of all material reproduced in this publication and apologize to copyright holders if permission to publish in this form has not been obtained. If any copyright material has not been acknowledged please write and let us know so we may rectify in any future reprint.

Except as permitted under U.S. Copyright Law, no part of this book may be reprinted, reproduced, transmitted, or utilized in any form by any electronic, mechanical, or other means, now known or hereafter invented, including photocopying, microfilming, and recording, or in any information storage or retrieval system, without written permission from the publishers.

For permission to photocopy or use material electronically from this work, access www.copyright.com or contact the Copyright Clearance Center, Inc. (CCC), 222 Rosewood Drive, Danvers, MA 01923, 978-750-8400. For works that are not available on CCC please contact mpkbookspermissions@tandf.co.uk

Trademark notice: Product or corporate names may be trademarks or registered trademarks and are used only for identification and explanation without intent to infringe.

ISBN: 978-1-032-49452-4 (hbk)
ISBN: 978-1-032-50285-4 (pbk)
ISBN: 978-1-003-39775-5 (ebk)

DOI: 10.1201/9781003397755

Typeset in Times New Roman
by MPS Limited, Dehradun

Contents

Section 1   Basics in Cell Culture

Goutham V. Ganesh, Karan Naresh Amin, Sundhar Mohandas,
R. Senthilkumar, and K.M. Ramkumar

Goutham V. Ganesh, Milan K.L., Sundhar Mohandas,
R. Senthilkumar, and K.M. Ramkumar

Goutham V. Ganesh, Kannan Hairthpriya,
Dhamodharan Umapathy, Md Enamul Hoque, and
K.M. Ramkumar

Section 2 Isolation and Analysis of Nucleic Acids and Proteins

Section 3   Cell Staining Techniques

Section 4   Cell Transfection and Single-Cell Analysis

Preface

This book is a practical manual designed for undergraduate and postgraduate students, scholars, teachers, and practitioners searching for a user-friendly text on the fundamentals of mammalian cell culture and techniques. The book presents a smooth transition from basics in cell culture to possible applications, including gene and protein expression analyses and subsequently to methods in genetic manipulation of cells, cellular staining techniques, and analyses of single cells. This book describes the principles underlying each technique and provides a detailed methodology to practice. The book will benefit both novices, to familiarize themselves with the basics of cell culture technique, and experts, to reacquaint or troubleshoot issues carrying out such methods.

This book is organized into four sections. The first section (Chapters 1–13) lays the foundational framework and concepts of mammalian cell culture and techniques, including the history of cell culture and its applications, cryopreservation for maintenance of cell lines, and aseptic techniques, including methods for detecting contamination in cell culture. The second section (Chapters 14–19) emphasizes the isolation and analysis of nucleic acids and proteins. The third section (Chapters 20–24) focuses on cell staining techniques, flow cytometric techniques, and bioplex assay. The fourth section (Chapters 25–27) focuses on cell transfection and single-cell analysis in depth.

Overall, this book thoroughly explains basic mammalian cell culture, techniques, and applications. The material is presented in a way that is adaptable to student use in formal courses as well as for experts. In addition to the fundamentals, attention is also given to modern applications and approaches to cell culture derivation, medium formulation, culture scale-up, and biotechnology, as presented by scientists who are pioneers in these areas. The sections include references to relevant Internet sites and other valuable sources of information. With this volume, it should be possible to establish and maintain a cell culture laboratory devoted to any of the many disciplines to which cell culture methodology is applicable.

K.M. Ramkumar
R. Senthilkumar
Md Enamul Hoque

Acknowledgments

First and foremost, I dedicate this book to my father, Mr. K.A. Mohanram, who has always been my idol and inspiration throughout my career. Your guidance and patience, no matter what I was doing, have helped develop me into who I am today. Thank you!

I express my deepest gratitude towards Professor K. Ramasamy, Former Director of Research at SRM Institute of Science and Technology, for his inspiring vision, motivation, and sustained efforts to see to the publication of this book.

I would like to express my special appreciation and thanks to my mentor Professor S. ThyagaRajan, Former Head, Department of Biotechnology, SRM Institute of Science and Technology for encouraging my research and allowing me to grow as a research scientist.

I would also like to thank Professor P. Rajaguru, Central University of Tamil Nadu, Thiruvarur, for his personal and professional guidance that taught me much about scientific research and work ethics.

I am especially indebted to Prof. Paulmurugan Ramasamy, Department of Radiology, Molecular Imaging Program at Stanford, Stanford University School of Medicine, who has supported my career goals.

This book would have been difficult to conceive or complete without the support and encouragement from Dr. Sarada, DVL, Associate Professor, Department of Biotechnology, SRM Institute of Science and Technology, Kattankulathur.

I feel pleased to have such dedicated guidance and extend my heartfelt thanks, especially to Prof. C. Muthamizhchelvan (Honourable Vice-Chancellor, SRMIST), Dr. S. Ponnusamy (Registrar, SRMIST), Dr. Gopal T.V (Dean, College of Engineering and Technology, SRMIST) and Dr. M. Vairamani (Dean, School of Bioengineering, SRMIST) for their goodwill and constant support in various forms.

I would like to thank my parents, whose love and guidance are with me in whatever I pursue. My special thanks to my brother, Mr. K.M. Krishnakumar, for his continuous support and encouragement.

I wish to thank my lab members for their hard work and contributions, which shaped the book into what it is.

I owe everything to God. I am truly grateful for his unconditional and endless love, mercy, and grace.

K.M. Ramkumar

Editor Biographies

Dr. K.M. Ramkumar is a professor of biotechnology at SRM Institute of Science and Technology (SRMIST), India. He received his doctoral degree from Bharathidasan University, Tiruchirapalli, India, where he was trained in cell signaling networks, particularly pancreatic β-cell apoptosis. He expanded his research into the cellular pathophysiology of diabetes during his post-doctoral fellowship at the Chonbuk National University Medical School, Republic of Korea. Further, during his research at the Stanford University School of Medicine, USA, he developed molecular imaging methods to study cellular and molecular pathways *in vitro* and *in vivo*.

During his academic tenure at SRMIST, Dr. Ramkumar established the Cell Culture Laboratory with a state of the art infrastructure. He has made several contributions to the cellular signaling network in diabetes. He has developed a high throughput cell-based assay for screening signaling molecules that restore cellular homeostasis. His research interests pertain to the identification of treatment targets for the management of diabetes. Much of his work has been on understanding the transcription factors involved in redox signaling associated with diabetes and its complications. He has also investigated the epigenetic regulation of cellular mediators involved in the pathophysiology of various diseases. Dr. Ramkumar has conducted several "Hands-on Training Workshops" related to cell culture.

He has over 150 scholarly contributions, including 100 peer-reviewed research papers, and based on his research accomplishments, Dr. Ramkumar has been ranked among the top 2% of scientists consecutively for two years, as per the global list released by Stanford University, USA. His research has received more the 3 Crores of funding from several Government agencies, including the Science and Engineering Research Board (SERB), Indian Council of Medical Research (ICMR), Department of Biotechnology (DBT), and Department of Science and Technology (DST) and Industries. He is also a recipient of various awards from Government agencies, including the "Young Scientist Award" from DST and the "SERB International Research Experience (SIRE)" award from SERB.

Dr. Ramkumar has established a research cohort with several hospitals and closely associates with clinicians, translating research from bench to bedside.

Dr. R. Senthilkumar is an associate professor/Ramalingaswami Re-entry Fellow [DBT] of the Department of Biotechnology, REVA University, Bangalore, India. He secured a PhD in biochemistry from Annamalai University, India. He did his three post-doctoral researches in anesthesiology at the University of Virginia, Charlottesville, Virginia, followed by the Department of Cancer Immunology & AIDS, Dana Farber Cancer Institute, Harvard University, Boston, Massachusetts, and the Department of Physiology and Biophysics, Virginia Commonwealth University, Richmond, Virginia, USA. Dr. Senthilkumar is a recipient of Rameshwardasji Birla Smarak Kosh Endowment award from Mumbai Medical Trust, Mumbai, India, and Ramalingaswami Re-entry Fellowship from the Department of Biotechnology, Ministry of Science and Technology, Government of India. He has more than 18 years of research experience in various fields such as alcoholic hepatotoxicity, voltage-gated calcium channels, protein kinase C phosphorylation, ADAM family, smooth muscle physiology, G-protein coupled receptor signaling pathways, and neurotransmitter transporters to the central nervous system. He has published nearly 75 research articles in various peer-reviewed international journals, 35 book chapters, 10 abstracts, and 65 conference proceedings/abstracts and attended various workshops/ seminars in India, Canada, and the USA. He has reviewed and screened the abstracts for National Biotechnology Conferences and the Annual Biomedical Research Conference for Minority Students (ABRCMS) and also served as special judge for the Biophysical Society award at the Virginia Piedmont Regional Science Fair. He conducts research in the area of G-protein coupled receptor mediated intracellular mechanisms by alcohol-induced digestive disorders in the gastrointestinal tract. Currently, he is mentoring graduate students and working on research projects from various funding agencies.

Prof. Dr. Md Enamul Hoque PhD (NUS, Singapore), PGCHE (Nottingham Uni, UK), FHEA (UK); FIMechE (UK), CEng (UK)

Dr. Md Enamul Hoque is a Professor in the Department of Biomedical Engineering at the Military Institute of Science and Technology (MIST), Dhaka, Bangladesh. Before joining MIST, he served in several leading positions at other global universities such as Head of the Department of Biomedical Engineering at King Faisal University (KFU), Saudi Arabia; Founding Head of Bioengineering Division, University of Nottingham Malaysia Campus (UNMC) and so on. He received his PhD in 2007 from the National University of Singapore (NUS), Singapore (1st in Asia and 8th in the World in QS World University Rankings 2024) with a globally prestigious scholarship from the Singaporean Government. He also obtained his PGCHE (Post Graduate Certificate in Higher Education) from the University of Nottingham, UK (18th in UK and 100th in QS World University Rankings 2024) in 2015. He is a Chartered Engineer (CEng) certified by the Engineering Council, UK; Fellow of the Institute of Mechanical Engineering (FIMechE), UK; Fellow of Higher Education Academy (FHEA), UK and Member, World Academy of Science, Engineering and Technology. To date, he has published 114 Journal Papers, 5 Special Issues in Journals, 11 Books, 75 Book Chapters, and 102 International Conference presentations/proceedings. As per Google Scholar citation report, his publications have attracted about 4000 citations with 30 h-index and 75 i-10 index. He received the Highest Publishing Scientist Award at MIST in March 2023, as well as the Outstanding Nano-scientist Award in the International Workshop on Recent Advances in Nanotechnology and Applications (RANA-2018) held in September 2018, AMET University, Chennai, India. His major areas of research interests include (but are not limited to) biomedical engineering, biomaterials, biocomposites, biopolymers, nanomaterials, nanotechnology, biomedical implants, rehabilitation engineering, rapid prototyping technology, 3D printing, stem cells and tissue Engineering. As per AD Scientific Index in 2023, he is the BEST Scientist (Biomedical Engineering) in the country (Bangladesh), 254 in Asia, and 1223 in the World. He is also listed among Top 2% Scientists in the World.

Contributors

Karan Naresh Amin
Department of Biotechnology
School of Bioengineering
SRM Institute of Science and
 Technology
Kattankulathur, Tamil Nadu, India

Goutham V. Ganesh
Department of Biotechnology
School of Bioengineering
SRM Institute of Science and
 Technology
Kattankulathur, Tamil Nadu, India

Kannan Harithpriya
Department of Biotechnology
School of Bioengineering
SRM Institute of Science and
 Technology
Kattankulathur, Tamil Nadu, India

Md Enamul Hoque
Department of Biomedical Engineering
Military Institute of Science and
 Technology
Dhaka, Bangladesh

Ravichandran Jayasuriya
Department of Biotechnology
School of Bioengineering
SRM Institute of Science and
 Technology
Kattankulathur, Tamil Nadu, India

Milan K.L.
Department of Biotechnology
School of Bioengineering
SRM Institute of Science and
 Technology
Kattankulathur, Tamil Nadu, India

Sundhar Mohandas
Department of Biotechnology
School of Bioengineering
SRM Institute of Science and
 Technology
Kattankulathur, Tamil Nadu, India

K.M. Ramkumar
Department of Biotechnology
School of Bioengineering
SRM Institute of Science and
 Technology
Kattankulathur, Tamil Nadu, India

R. Senthilkumar
Department of Biotechnology
School of Applied Sciences
REVA University
Bangalore, Karnataka, India

Dhamodharan Umapathy
Department of Biotechnology
School of Bioengineering
SRM Institute of Science and
 Technology
Kattankulathur, Tamil Nadu, India
and
Department of Research
Adhiparasakthi Dental College and
 Hospital
Melmaruvathur, Tamil Nadu, India

Foreword A and B

DEPARTMENT OF RADIOLOGY

FOREWORD A

I am pleased to write a foreword for the book entitled ***Advanced Mammalian Cell Culture Techniques: Principles and Practices*** edited by *Prof. K.M. Ramkumar, Dr. R. Senthilkumar, and Prof. Dr. Md Enamul Hoque*, and published by CRC Press, Taylor and Francis Group, USA. The editors, with their extensive expertise in the cell culture systems and their various applications, are well qualified to compile this book, with several chapters addressing stepwise protocol and background information. Similarly, the selected contributing authors have firsthand experience in primary cells, mammalian cell lines, advanced cell culture techniques, and other fields of cell imaging that are evident from their reputed publications. This book also comprehensively and succinctly discusses the isolation and analysis of nucleic acids, proteins, cell staining techniques, cell transfection, and single cell analysis.

I am currently a professor in the Department of Radiology at Stanford University and a member of the Canary Center for cancer early detection, the Molecular Imaging Program at Stanford (MIPS), and the Stanford BioX program. I have more than 30 years of research experience in molecular and cellular biology, and 20 years of experience in the use of *in vivo* imaging modalities (bioluminescence, fluorescence, PET, SPECT, US, MRI, and CT) for monitoring different cellular events in mice and rats. My research group focuses on developing new imaging assays for studying cellular signal transduction networks in cancer and other diseases. Specifically, we apply our extensive experience in molecular biology to develop *in vivo* imaging assays that can be used for monitoring basic cellular processes and post-translational modifications of proteins, such as methylation, phosphorylation, sumoylation, and many more. We have developed split-reporter protein complementation systems for various reporter proteins (luciferases, fluorescent proteins, and thymidine kinase), and use them for designing various sensors to study cellular signaling processes. Some of the main applications of these assays include imaging the tumor microenvironment, as well as immune cytokine signaling in cancer and infectious diseases. Other applications where we are currently applying these assays include studying protein–protein interactions involved in estrogen receptor signaling, Nrf2-mediated antioxidant signaling in chemoresistance, p53-sumoylation mediated chemotherapy responses in cancer, NFkB-mediated cytokine signaling in cancer, and signaling mechanisms associated with APP and Tau protein sumoylations in Alzheimer's disease.

In cancer therapy, we are establishing microRNA-based reprogramming approaches to sensitizing drug-resistant cancers (breast cancer, hepatocellular carcinoma, and glioma) to commonly used chemotherapies. We mainly target oncogenic and tumor suppressor microRNAs (miR-21, miR-10b, miR-122, and miR-100) that are dysregulated in cancers to improve cancer therapy. We have shown tremendous progress in this area of research, with many publications in high-impact journals. In synthetic biology, we recently invented the application of a high-pressure microfluidic system in the reconstruction of biomolecules derived from cells (proteins and lipids) along with synthetic sources (phospholipids, polymers, and surfactants) to develop self-assembled nano- and micro-structures that mimic biological membranes for drug delivery applications. As part of this process, we developed biomimetic microbubbles (biMBs) and nanobubbles (biNBs) using tumor cell derived exosomes (TDEs) for cancer immunotherapy applications. We also developed efficient way to reconstruct microvesicles (MVs) derived from cell sources into EVs of uniform sizes and load miRNAs using a microfluidic-based system.

The continuing development of new and improved techniques in mammalian cell culture should match with the changes in other fields. Of course, one would expect to have all the "usual" chapters be present, and at the same time readers should not be disappointed. This book covers all the techniques related to mammalian cell culture, and it is difficult to see what might be missing. I certainly could not find anything important that had been omitted. The images are well-chosen and superbly reproduced. This book also addresses the basic information about different cells, how to maintain the cells, preparation of media, expected contaminations, and how to detect and eliminate them, aseptic cell culture, and essentials of cell culture.

With the help of the chapters from different contributors drawn from India and Bangladesh, the editors have compiled a textbook that is not only the first of its kind, but one that I would rate as the most informed. It is one of the clearest and most up-to-date books in cell culture techniques currently available in world literature. The editors divided the book into 4 sections with 27 detailed chapters, each of which is well-drafted, clearly written, and illustrated with comprehensive bibliography. This book will have a wide appeal and will be helpful for those who are working in many scientific and medical disciplines. Graduate and undergraduate students will benefit from the breadth of knowledge and bibliography provided in each chapter to inform them of the basic mechanisms as they relate to mammalian cell culture and other related techniques.

Sincerely,

 **Ramasamy Paulmurugan**, PhD, Professor of Radiology, Molecular Imaging Program at Stanford (MIPS), Canary Center for Cancer Early Detection, Department of Radiology, Stanford University School of Medicine, Palo Alto, CA.

FOREWORD B

It is my great pleasure to provide the foreword for the book entitled *Advanced Mammalian Cell Culture Techniques: Principles and Practices*. Despite the occasional appearance of thoughtful works devoted to elementary or advanced cell culture methodology, a place remains for a comprehensive and definitive volume that can be used to advantage by both novices and experts in the field. In this book, *K.M. Ramkumar, R. Senthilkumar, and Md Enamul Hoque* present the relevant methodology within a conceptual framework of cell biology, genetics, and physiology that renders technical cell culture information in a comprehensive, logical format. This allows topics to be presented with an emphasis on troubleshooting problems from a basis of understanding the underlying theory. The material is presented in a way adaptable to student use in formal courses. Also, it is functional when used daily by professional cell culture technicians in academia and industry. In addition to the fundamentals, attention is also given to modern applications and approaches to cell culture derivation, medium formulation, culture scale-up, and biotechnology, presented by scientists who are pioneers in these areas. With this volume, it should be possible to establish and maintain a cell culture laboratory devoted to any of the many disciplines to which cell culture methodology is applicable. The volume includes references to relevant sites and other valuable sources of information. The extraordinary effort required to write a book that comprehensively and authoritatively presents the state of knowledge in an arena as broad and ever-expanding as mammalian cell culture and its applications is a daunting task. I am certain that the readers, including faculties, researchers, and students, will find this book extremely informative, interesting, and inspiring.

I, Dr. Gautam Sethi, am currently working as a tenured associate professor in the Department of Pharmacology, Yong Loo Lin School of Medicine, National University of Singapore. The focus of my research over the past few years has been to elucidate the mechanism(s) of activation of oncogenic transcription factors by carcinogens and inflammatory agents and the identification of novel inhibitors of these proteins for the prevention and treatment of cancer. The findings of our research work have so far resulted in more than 400 scientific publications in high-impact factor peer-reviewed journals and several international awards.

I have referred to a large number of books published on various aspects of the beneficial usage of cell lines and have made an effort to unify all the content of scattered research literature in the area of mammalian cell culture. This book, however, is not just a collection of chapters but an essence of the diverse cell culture and cell line techniques and their applications. In all chapters, the authors have provided the basic information relevant to the topics, and at the same time, they have described the perspective knowledge about using high throughput technological approaches. The editors bring an understanding of the milieu of clinical research, experimental research, and pharmacology which, revenue collected, focuses on the most active areas of cell culture techniques. This book could be an informative resource, in the form of a condensed handbook, for research students as well as advanced researchers.

Sincerely,

Dr. Gautam Sethi, Associate Professor, Department of Pharmacology, Yong Loo Lin School of Medicine, National University of Singapore, Singapore.

Section 1

Basics in Cell Culture

1 Highlights

History of Cell Culture and Its Applications

Goutham V. Ganesh, Karan Naresh Amin, and Sundhar Mohandas
Department of Biotechnology, School of Bioengineering, SRM Institute of Science and Technology, Kattankulathur, Tamil Nadu, India

R. Senthilkumar
Department of Biotechnology, School of Applied Sciences, REVA University, Bangalore, Karnataka, India

K.M. Ramkumar
Department of Biotechnology, School of Bioengineering, SRM Institute of Science and Technology, Kattankulathur, Tamil Nadu, India

1.1 MILESTONES IN CELL CULTURE TECHNIQUES

The earliest attempts in cell culture were mainly based on tissue culture or explants cultured to observe cellular growth from explant tissue radially into the underlying substratum. Maintenance of chicken embryos in artificial fluids by Willhelm Roux in 1885 and culture of frog embryonic nerve fibers was successful. However, these early days witnessed simple experimental set-ups in which aggregated pieces of tissue were grown under laboratory conditions. They were quite unlike the culture of suspension of cells that we are familiar with today. With the advent of cell strains, or clones of cells derived from mouse L929 fibroblasts [1], and the propagation of HeLa cells (isolated by George Gey from Johns Hopkins University Hospital in 1952), cell culture began to be taken seriously, and cell culture laboratories around the world institutionalized materials in cell culture techniques and policies concerning obtaining human tissue samples in this field of study. Similarly, the development of mutant cell lines using Chinese hamster ovary cells enabled researchers to study specific pathways in cellular metabolism that otherwise would not have been possible at that time [2].

DOI: 10.1201/9781003397755-2

The development of monoclonal antibodies and related cell culture-based products hastened the development of cell culture applications for industrial-scale production [3], including therapeutic applications, diagnosis, and, more importantly, further research to foster newer treatments. The use of transformed or immortalized cell lines, as opposed to primary cells, which have a finite life span, in various cell culture applications, such as with hybridoma technology in monoclonal antibody production mentioned above, was considered advantageous due to their homogenous genetic background, easier availability (unlike primary cells, which necessitated euthanizing animals), among other reasons. However, unlike transformed cells (derived from tumors), other immortalized cell lines are derived from normal cells transduced adeno- or lentiviruses commonly to create transient or stable cell lines. Of late, developments regarding conditional cell immortalization wherein the addition of a substrate, for instance, resulted in cell immortalization that otherwise, or when the substrate was removed, reverted cells back to their normal state [4].

In 1993, the Food and Drug Administration (FDA) revised the points to consider in the characterization of cell lines for the production of biologics by manufacturers. Such topics included the history of the cell line to be used in production, generation of cell banks and management of cell cultures, and quality control testing for micro-organisms and tumorigenicity. One important facet in managing cell cultures is the increasing use of synthetic media in industrial or commercial applications. However, a greater emphasis is required in this direction to minimize batch-to-batch variation in production and optimization of conditions for cells cultured in serum-free media [5]. Another exciting area of research that is equally important in both basic research and applied fields of pharmaceutical analysis and precision medicine is the use of 3D scaffolds that closely mimics the in vivo situation for the culture of tumor isolates from clinical samples in cell-free systems that can be scaled-up for commercial use [6]. The use of xeno-free culture systems is amenable for use with the culture of human embryonic stem cells for cell-based therapeutic applications [7].

The diverse applications of cell culture systems have found areas of research with enormous clinical implications in 3D printing of tissues and, very importantly, organs, and also in organ preservation via supercooling, which are vital in addressing existing shortcomings in transplantation. Successful engineering of nose cartilage tissue by freeform reversible embedding of hydrogel [8] and a novel 3D bioprinting of diabetic human skin with structural similarities to human disease skin [9] are important milestones that hasten our fundamental understanding of pathology and devise novel treatments. Sub-zero organ preservation is an approach to extending the viability of organs to be transplanted, and a supercooling methodology was demonstrated to improve the *ex vivo* life of human livers at the time of transplantation [10].

REFERENCES

[1] Sanford, K.K., Earle, W.R., & Likely, G.D. (1948). The growth *in vitro* of single isolated tissue cells. *Journal of National Cancer Institute*, 9, 229.
[2] Kao, F., Chasin, L., & Puck, T.T. (1969). Genetics of somatic mammalian cells, X. Complementation analysis of glycine-requiring mutants. *Proceedings of the National Academy of Sciences of the United States of America*, 64(4), 1284–1291.

[3] Shukla, A.A. & Thommes, J. 2010). Recent advances in large-scale production of monoclonal antibodies and related proteins.*Trends in Biotechnology, 28*(5), 253–261. 10.1016/j.tibtech.2010.02.001

[4] Wall, I., Toledo, G.S., & Jat, P. (2016). Recent advances in conditional cell immortalization technology. *Cell Gene Therapy Insights*, 2(3), 339–355. 10.18609/cgti.2016.044

[5] Yao, T. & Asayama, Y. (2017). Animal-cell culture media: History, characteristics, and current issues. *Reproductive Medicine and Biology*, 16, 99–117. 10.1002/rmb2.12024

[6] Jordahl et al. (2019). Engineered fibrillar fibronectin networks as three-dimensional tissue scaffolds. *Advanced Materials*, 1904580. 10.1002/adma.201904580

[7] Desai et al. (2015). Human embryonic stem cell cultivation: Historical perspective and evolution of xeno-free culture systems. *Reproductive Biology and Endocrinology*, 13, 9. 10.1186/s12958-015-0005-4

[8] Lan, X., Yan, L., Erkut, E.J.N., Kunze, M., Mulet-Sierra, A., Gong, T., Osswald, Martin., Ansari, K., Seikaly, H., Boluk, Y., Adesida, A.B. (2021). Bioprinting of human nasoseptal chondrocytes-laden collagen hydrogel for cartilage tissue engineering. FASEB Journal,35(3),e21191 10.1096/fj.202002081R

[9] Kim, B.S., Ahn, M., Cho, W.W., Gao, G., Jang, J., & Cho, D.W. (2021). Engineering of diseased human skin equivalent using 3D cell printing for representing patho-physiological hallmarks of type 2 diabetes *in vitro*. *Biomaterials*. 10.1016/j.biomaterials.2021.120776

[10] de Vries, R.J., Tessier, S.N., Banik, P.D. et al. (2019). Supercooling extends preservation time of human livers. *Nature Biotechnology*, 37, 1131–1136. 10.1038/s41587-019-0223-y

2 Aseptic Technique and Essentials in Cell Culture

Goutham V. Ganesh, Milan K.L., and Sundhar Mohandas
Department of Biotechnology, School of Bioengineering, SRM Institute of Science and Technology, Kattankulathur, Tamil Nadu, India

R. Senthilkumar
Department of Biotechnology, School of Applied Sciences, REVA University, Bangalore, Karnataka, India

K.M. Ramkumar
Department of Biotechnology, School of Bioengineering, SRM Institute of Science and Technology, Kattankulathur, Tamil Nadu, India

2.1 ASEPTIC TECHNIQUE

The importance of aseptic handling of cell cultures, primary or otherwise, cannot be overstated in cell culture, and several problems can arise, including microbial- and cross-contamination with other cell cultures, cell line misidentification, and others that can pose a serious threat to both basic research and products intended for clinical/therapeutic needs. *Aseptic technique* is a term that broadly covers a set of guidelines or instructions that pertain to methods to which sterile conditions need to be strictly adhered to [1]. The most critical aspects for the successful implementation of asepsis include (i) the source material, including cells, culture medium, and other biological reagents, (ii) equipment such as cell culture plastic- and glassware, instruments, and other devices that come in contact with cells, and more importantly, (iii) laboratory staff.

A step-by-step method for aseptic use of a level II biosafety cabinet (BSC, illustrated in Figure 2.1), common in cell culture laboratories and classified according to the level of containment required (level I, II, and III), will be described below. The class II BSCs are characterized by a unidirectional and vertical flow of air with access to the front of the cabinet [2].

DOI: 10.1201/9781003397755-3

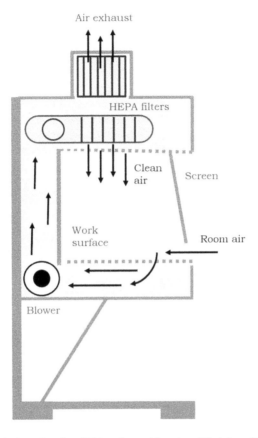

FIGURE 2.1 Rendering of a class II biosafety cabinet (modified from Rosalie J. Coté [2]). Note that the arrows marked in black represent clean air entering the work area. Exhaust air leaves the work area under the work surface and either leads to the air exhaust outside or is recycled back in after filtering through HEPA filters and the top of the cabinet.

2.2 MATERIALS

- Alcohol 70% (v/v) in sterile water
- Sodium hypochlorite
- Personal protective equipment, or PPE (sterile gloves, lab coat, safety visor, overshoes, headband)
- Biosafety level cabinet at a level of containment that is appropriate

2.3 PROCEDURE

1. Before use, empty and clean the BSC floor with 70% alcohol, after which clean and place all non-biological reagents, cell culture plastics, and other equipment in the cabinet for surface-level decontamination using UV

light for about 20–30 minutes. Meanwhile, fresh lab coats, gloves, and other disposable personal wear must be available.

2. Spray gloves with 70% alcohol, and before beginning work, place all materials and equipment in the cabinet for sterilization by UV light for about 20–30 minutes.

3. Ensure no air vents inside the cabinet are blocked and while clean, unused equipment is kept in one corner, provision is made at the other end for plastic and liquid discard and putting away other items after use. Plan to keep the floor area for a clean-to-dirty workflow.

4. Before using cabinet equipment (media bottles, pipettes, pipette tips, pipette aids), wipe them with the tissue soaked in 70% alcohol.

5. Move sterile biological liquids (after pre-warming in a water bath to around 37°C) into the cabinet after thoroughly cleaning with 70% alcohol.

6. During work, refrain from touching anything outside the cabinet (especially your face and hair) to prevent gloves contamination. Re-spray gloves with 70% alcohol as above if they become contaminated.

7. Finally, bring in cells cultured in appropriate flasks or dishes and ensure careful handling of cultures with deliberate movements and not touching non-sterile flasks with pipettes or items that directly come in contact with the culture medium or cells.

8. Make sure not to move hands directly over open culture vessels or flasks and not touch the necks of flasks or mouths of sterile tubes, even with gloved hands.

9. Finally, at the end of sterile operations, decontaminate the BSC floor again and remove used plasticware and liquids appropriately to waste bins. Before removing any equipment or material from the cabinet after finishing work, spray the work surfaces with 70% alcohol and wipe dry.

10. All biological waste, including cell material that can shed viral DNA, should be dispensed in appropriately designated bags for disposal by a concerned agency.

11. It is important to move slowly in and out of the cabinet so that the air within can circulate properly.

12. It is a good idea to frequently wipe down the cabinet surfaces with a disinfectant and fumigate the cabinet according to the manufacturer's instructions. However, you must ensure that the laboratory space is safe to fumigate and consult your on-site Health and Safety Advisor.

In addition to a BSC, cell culture laboratories also, at a minimum, require incubators (to provide a humidified atmosphere with a temperature maintained at 37°C, and in addition, a constant supply of 5% CO_2 piped into the incubator), a centrifuge, sterilization equipment such as the autoclave (for moist heat sterilization wherein cell culture liquids such as PBS can be sterilized at 121 psi for 15–20 minutes at 121°C) and a hot air oven for dry heat sterilization of pipettes and other glassware, an inverted microscope to monitor cells for growth, proliferation, and visually inspect contamination of cell cultures by micro-organisms. It is preferable to use

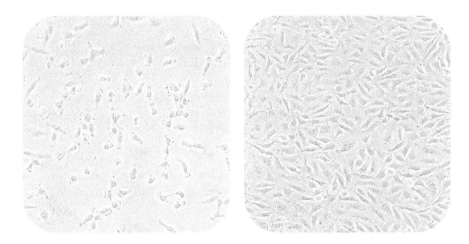

FIGURE 2.2 Light microscopic images (10×) of low- (left image) and high- (right image) density cultures of EA. hy926 endothelial cells.

incubators that can be programmed for sterilization of the entire incubator periodically. A general guideline for preparing and culturing an immortalized, anchorage-dependent cell line, endothelial cell line EA. hy926, is given in Table 2.1 and Figure 2.2.

TABLE 2.1

Cell Culture Seeding and Harvest Cell Densities for Different Cell Culture Vessels

Cell Culture Vessel	Growth Surface Area (cm²)	Seeding Density	Confluent Density	Recommended Well Volume (mL)
96 well plate (flat bottomed)	0.32	1×10^4	0.35×10^4	0.1–0.2
24 well plate	1.9	5×10^4	2×10^5	0.5–1
12 well plate	3.8	1×10^5	5×10^5	1–2
6 well plate	9.6	$4–6 \times 10^5$	$1–1.5 \times 10^5$	1–2.5
T-25 flask[#]	25	1.25×10^5	$2–3 \times 10^6$	4–6
T-75 flask[#]	75	3.75×10^5	$8–10 \times 10^6$	10–20

Unless otherwise indicated, the numbers are suitable for seeding and recovery in experimental assays [3]. Cell confluent densities vary according to cell type and culture conditions. The reported numbers are meant for EA. hy926 cells. [#]for routine culture.

2.3.1 CELL LINE BACKGROUND

Cell type	Human umbilical vein cell line
Tissue origin	Human umbilical vein
Biosafety containment	BSL-1
Immortalization	Hybridoma technology using HAT selection
Recommended medium	DMEM with 10% FBS, at a final concentration (v/v)

2.3.2 ESSENTIAL INFORMATION, INCLUDING CELL NUMBERS

Typical confluent density	2–3 million cells per 25 cm^2 surface area
Seeding density	1.25×10^5 cells per 25 cm^2 flask (minimum requirement)
Medium renewal	Alternate days but varies with cell confluency
Cryo-preservation medium	Complete growth medium supplemented with 5% DMSO

REFERENCES

[1] R.I. Freshney (2007). *Culture of animal cells, A manual of basic technique and specialized applications*, 6th edition. John Wiley & Sons, Inc., U.S., 10.1002/9780470649367

[2] Rosalie J. Coté (1998). Current methodologies in cell biology 1.3.1–1.3.10. 10.1002/0471143030.cb0103s00

[3] https://www.thermofisher.com/in/en/home/references/gibco-cell-culture-basics/cell-culture-protocols/cell-culture-useful-numbers.html

3 Laboratory Safety

Goutham V. Ganesh and Kannan Harithpriya
Department of Biotechnology, School of Bioengineering,
SRM Institute of Science and Technology, Kattankulathur,
Tamil Nadu, India

Dhamodharan Umapathy
Department of Biotechnology, School of Bioengineering,
SRM Institute of Science and Technology, Kattankulathur,
Tamil Nadu, India

Department of Research, Adhiparasakthi Dental College and
Hospital, Melmaruvathur, Tamil Nadu, India

Md Enamul Hoque
Department of Biomedical Engineering, Military Institute of
Science and Technology, Dhaka, Bangladesh

K.M. Ramkumar
Department of Biotechnology, School of Bioengineering,
SRM Institute of Science and Technology, Kattankulathur,
Tamil Nadu, India

3.1 KEY ISSUES IN CELL CULTURE LABORATORY

The setting-up of laboratory safety methodologies is paramount to avoid, or at the very least, minimize several hazards that could happen in the cell culture laboratory. In addition to the lab being routinely used by scientifically trained personnel, service staff and new members utilize the space momentarily or otherwise for entirely different reasons. It is essential that new lab members, their seniority notwithstanding, are cognizant and equally compliant with the lab safety procedures to avoid unwanted mishaps occurring in the cell culture laboratory that could pose risks to themselves and other personnel as well as cell cultures. Following are some of the key issues concerning safety in a cell culture laboratory.

3.1.1 BIOSAFETY LEVEL CONTAINMENT AND BIOHAZARDS

The level of biosafety required for a particular experiment depends on the source of biomaterial and the procedure being conducted in that particular experiment. The source of biomaterial may be non-pathogenic, potentially pathogenic, or hazardous

DOI: 10.1201/9781003397755-4

TABLE 3.1

Levels of Containment for Handling and Culture of Biological Materials from Different Sources

Procedure to be Undertaken	Level of Containment Required	Equipment Needed
Preparation of reagents, including PBS and cell culture media	Good laboratory practice	Adherence to standard microbiological work on an open bench; optional: Class I or II biosafety cabinet
Culture of cells emanating from non-human and non-primate species	1	Class I biosafety cabinet
Primary and secondary cultures of cells from humans and other primates	2	Class II biosafety cabinet
Transfected human cells, the culture of human and animal tumor cells	3	Class II biosafety cabinet
Handling cell cultures with retroviral constructs	3	Class II biosafety cabinet
Samples were taken from human tissue containing known pathogens	4	Class III biosafety cabinet with a heightened level of containment*

Note

* It must be noted that the Class III biosafety cabinet features higher containment levels, including gloved pockets for handling cultures. In addition to other precautions, the exhaust air vent will be equipped with a pathogen trap to minimize the occurrence of pathogen escape from the containment facility.

(the biomaterial source, such as the cell line, may be virally transformed and still secrete viral products that may be pathogenic to human beings and animals alike or in the context of the procedure being adopted, such as the use of a cytotoxic drug or a radioisotope) or otherwise pathogenic, which requires the highest level of containment. An illustration of this is provided in Table 3.1.

3.1.2 Acute Toxicity from Accidental Exposure to Toxic Chemicals, Gases, and Liquids, and Physical Injury

Biosafety guidelines are generally regulated by the institution and enforced daily by the concerned department and/or the Principal Investigator. In addition to personnel training in handling lab safety materials, utmost care must be taken to ensure proper working and maintenance of lab equipment and checking for electrical faults and disaster management procedures. The essential health and safety manual [1] provides a detailed account of guidelines to be followed in the research laboratory. In addition to general injuries arising from broken glassware and sharp instruments, exposure to toxic chemicals and solvents that are otherwise nontoxic, such as DMSO, but that can easily penetrate gloves and the skin barrier to expose the user to toxic drugs [2], tissue samples containing pathogens, inflammable gases such as oxygen, and the use of radiological isotopes can cause damage to users, bystanders, and cell cultures alike. Other gases such as CO_2, though nontoxic, can cause asphyxiation, and, it must be said liquid nitrogen, used commonly in cell culture laboratories for cryopreservation of cells and tissues, poses a considerable danger to handling personnel due to the danger of (i) explosion from liquid nitrogen leaking from vials or (ii) frostbite upon exposure to unprotected or inadequately covered skin. Needle injuries are a particular concern, especially when handling blood or tissues sampled from patients suffering from diseases such as HIV.

Assessment of risks in the laboratory needs to be properly carried out, and materials that pose considerable risks should be handled according to existing standard operating procedures (SOPs), including various stages in preparation, execution, and termination of the methodology deemed as a risk in the laboratory. The Health and Safety website constantly updates relevant information, including the risks relating to the recent COVID-19 outbreak disease outbreak in laboratories, advice on genetically modified organisms (GMOs), specific animal pathogens encountered when handling human or animal cell or tissue samples, and other related information. It must be kept in mind that careful assessment of the nature and degree of severity of a specific risk or hazard is essential. The overall surrounding conditions also play a major role in accounting for the risk. Equal attention must be paid in regards to the disposal of hazardous waste, including biological material, and adequate provisions must be made to sterilize the equipment and laboratory after conducting such procedures, including the use of a strong disinfectant such as sodium hypochlorite solution (bleach) and fumigation with either formaldehyde gas or hydrogen peroxide.

3.1.3 Treatment and Disposal of Cell Culture Waste

Cell culture laboratories are required to follow certain guidelines in regards to handling and disposal of cell culture wastes depending on the type of biomaterials (non-pathogenic or otherwise to users of the facility) and their potential to cause problems due to accidental release or, if untreated or improperly treated, biowaste that is disposed of outside. Most cell lines require a BSL 2 biosafety cabinet, and laboratories that work with infectious agents, such as cells derived from HIV patients, require a higher level containment, such as level 3 [3]. The most common risks in handling biomaterials that can cause harm include the biological agents or toxins that can be transmitted via ambient air; therefore, air ventilation mechanisms are necessary to regulate unidirectional air flow with exhaust air filters in place that decontaminate the air circulated through the work area that can then be recycled back or released outside.

Accidental direct exposure to biological material is the other frequently encountered risk to personnel. Therefore, emphasis on personal protective equipment (PPE), such as gloves, eyewear, and overcoats, and regulations to treat accidental spills and decontaminate biological wastes using autoclaves or by chemical means (using hypochlorite solution or bleach) are recommended. A bio-decontamination waste disposal mechanism involving heat inactivation was published to handle both biological materials and wastes emanating from handling such biomaterials [4]. The most common biowaste from a cell culture laboratory is the spent cell culture medium containing antibiotics that is of environmental concern if disposed of without necessary prior treatment, such as autoclaving and/or removing such media marked "chemical waste" to ensure proper handling and disposal [5].

3.1.4 Dangers in Liquid Nitrogen Handling in the Lab

The use of liquid nitrogen in the cryopreservation of cells is well documented and goes back in history, spanning more than a few decades [6]. Freezing and reviving cells are essential in cell culture for their varied clinical applications [7]. The risks accompanying handling liquid nitrogen are very real. The inherent dangers associated with nitrogen being odorless and colorless (accumulation of nitrogen can potentially lead to asphyxiation if undetected), and the extremely low temperatures in using liquid nitrogen (at −196°C) expose the user to injury from thermal and explosive injuries when handling cryo-vials recovered from liquid nitrogen containers [8]. Diligent care must therefore be ensured, including implementation and adherence to protocols for appropriate storage, maintenance, and monitoring of liquid nitrogen storage vessels, and training staff to ensure their safety while recovering cryo-stocks from liquid nitrogen storage is crucial importance.

REFERENCES

[1] OSHA, The essential health & safety manual. 2009.
[2] Ostertag, W. and I.B. Pragnell, Changes in genome composition of the Friend virus complex in erythroleukemia cells during the course of differentiation induced by dimethyl sulfoxide. *Proc Natl Acad Sci U S A*, 1978. **75**(7): pp. 3278–3282.
[3] Mourya, D.T., et al., Establishment of Biosafety Level-3 (BSL-3) laboratory: important criteria to consider while designing, constructing, commissioning & operating the facility in Indian setting. *Indian J Med Res*, 2014. **140**(2): pp. 171–183.
[4] Gregoriades, N., et al., Heat inactivation of mammalian cell cultures for biowaste kill system design. *Biotechnol Prog*, 2003. **19**(1): pp. 14–20.
[5] Meyer, E.L., G. Golston, S. Thomaston, M. Thompson, K. Rengarajan, and P. Olinger, Is Your Institution Disposing of Culture Media Containing Antibiotics? *Applied Biosafety*, 2017. **22**(4): p. 4.
[6] Nagington, J. and R.I. Greaves, Preservation of tissue culture cells with liquid nitrogen. *Nature*, 1962. **194**: pp. 993–994.
[7] Jang, T.H., et al., Cryopreservation and its clinical applications. *Integr Med Res*, 2017. **6**(1): pp. 12–18.
[8] Sigma-Aldrich, Fundamental Techniques in Cell Culture Laboratory Handbook. 2015.

4 Preparation of Media for Cell Culture

Milan K.L., Goutham V. Ganesh, and Sundhar Mohandas
Department of Biotechnology, School of Bioengineering, SRM Institute of Science and Technology, Kattankulathur, Tamil Nadu, India

Md Enamul Hoque
Department of Biomedical Engineering, Military Institute of Science and Technology, Dhaka, Bangladesh

K.M. Ramkumar
Department of Biotechnology, School of Bioengineering, SRM Institute of Science and Technology, Kattankulathur, Tamil Nadu, India

4.1 INTRODUCTION

Molecular interactions *in vivo* that aid in tissue or organ growth and development are intricate processes that cannot be replicated outside the body. To ensure the steady growth and proliferation of cells in culture, proper requirements need to be provided. One critical aspect is the formulation of media used in cell cultures.

The presence of tissue and systemic interactions that generally occur in the body of an organism can't support the growth of cultured cells. Therefore, the needs of cell cultures differ from those of animal cells. For cells to divide rapidly in culture, they need an abundance of easy-to-use nutrients. Media formulations for cell culture are meant to mimic the conditions of nature to the greatest extent possible but with additional nutrients and building blocks for cellular growth. Basal media contains nutrients, vitamins, and minerals, and in addition, the animal serum contains additional nutrients, growth factors, and hormones [1]. Considering that the exact components of the supplemental serum are not known, this type of media is called "undefined media." Biological serum can be obtained from various animals, including calves, bovine fetuses, horses, or humans.

One can choose from a variety of formulations of basal media, including Eagle's Minimal Essential Medium (MEM), RPMI 1640, or Dulbecco's Modification of Eagle's Medium (DMEM).

Different cell types prefer different formulations to grow optimally. Despite the slight differences in the exact recipes, all basal media consist of the following elements:

DOI: 10.1201/9781003397755-5

- A carbon source (glucose/glutamine) is the source of energy.
- Amino acids are essential for the synthesis of proteins.
- Vitamins assist in cell growth and survival.
- A balanced salt solution provides cofactors for enzyme reactions, cell adhesion, etc., this solution is isotonic.
- Phenol Red is a non-nutritional dye used as a pH indicator. In higher pH environments, Phenol Red changes from orange/red to yellow at pH 7–7.4 and from purple to yellow at acidic pH levels below pH 7.4. pH can be monitored by observing the color change of the media in question.
- A pH-balanced media is maintained by adding bicarbonate or HEPES buffers.

In preparation for "complete media," serum is added to the basal media. A complete media can be added with antibiotics to prevent the growth of bacteria. It is not recommended to use antibiotics continuously since it encourages the growth of more aggressive and resistant bacteria [2].

The basal media can also be supplemented with growth factors, proteins, and other components without serum to prepare the serum-free media. Since the entire composition of the media can be identified, this type of media is referred to as "defined media." It is possible to customize and select a defined media for a particular cell type or laboratory condition. The preparation can take longer using defined media since the components must be added individually in the correct concentrations [3]. It is also possible to buy commercially available components usually present in serum separately in preparation for serum-free media.

4.2 MATERIALS

- Serological pipette
- Sterilized 50 ml tubes
- 450 ml bottle of basal media. The majority of manufacturers include 450 ml of basal media in each bottle to simplify the preparation process.
- 200 mM L-glutamine (optional)
- Fetal-bovine serum
- Antibiotics at 100X concentration [Penicillin and Streptomycin (Pen-strep)] (Optional)

4.3 METHODOLOGY

Undefined media containing fetal-bovine serum will be prepared for the cells. To prepare 500 ml of complete media, follow these steps:

1. Wash your hands before you start.
2. Clean the laminar hood with 70% ethanol
3. In the 37°C water bath, defrost the fetal-bovine serum, L-Glutamine, and Pen-strep bottles stored in the freezer. After wiping the bottles with alcohol, swirl them a few times and then open them in the hood.

4. Remove the basal media bottle's cap. Using a serological pipette, add 50 ml of fetal bovine serum to the 450 ml media to arrive at a final FBS concentration of 10%.
5. Add 5 ml of L-Glutamine to the bottle using a new pipette, so the final concentration of L-Glutamine is 2 mM. If L-Glutamine is already included in the basal media, extra L-Glutamine is not required. There may be differences in L-Glutamine concentrations between various media used for different cell lines (0.5–10 mM).
6. Add 5 ml of 100X antibiotics if needed to the prepared 500 mL medium with FBS.
   a. (pen-strep to the bottle. Every time you change solutions, use a new serological pipette.)
7. Recap all bottles.
8. Swirl the bottle to mix the contents.
9. Media must be prepared using sterile reagents. (You can ensure the sterility of the freshly prepared complete media by filtering it through a 0.2 μm filter.)
10. Aliquot the complete media into sterile plastic tubes of 50 ml, each using a new serological pipette.
11. Label each tube of complete media with the date and initials. Also, note if antibiotics were used.
12. Be sure to cap the tubes and store them in the refrigerator. Use one tube at a time. All media can be stored in the fridge for up to a month.

4.4 KEY POINTS

1. Media must be prepared using sterile reagents. (You can ensure the sterility of the freshly prepared complete media by filtering it through a 0.2 μm filter.)
2. Each tube of complete media should be labeled with the date and initials. Also, note if antibiotics were used.
3. Be sure to cap the tubes and store them in the refrigerator. Use one tube at a time. All media can be stored in the refrigerator for up to a month.

REFERENCES

[1] Freshney, R.I. (2005) *Culture of animal cells: a manual of basic techniques.* 5th ed. Hoboken, New Jersey. John Wiley & sons Inc.
[2] Afshar, G. et al. (2018) Basics of Cell Culture: A student laboratory manual. Volume 7. Edited by: Deborah Blancas, Carin Zimmerman Ph.D., and Edie Kaeuper Ph.D. (2011)
[3] Segeritz, C.P., Vallier, L. et al. (2017) Cell Culture: Growing Cells as Model Systems In Vitro. *Basic Science Methods for Clinical Researchers.* 10.1016/B978-0-12-803 077-6.00009-6.

5 Primary Cell Culture

Milan K.L. and Goutham V. Ganesh
Department of Biotechnology, School of Bioengineering,
SRM Institute of Science and Technology, Kattankulathur,
Tamil Nadu, India

Dhamodharan Umapathy
Department of Biotechnology, School of Bioengineering,
SRM Institute of Science and Technology, Kattankulathur,
Tamil Nadu, India

Department of Research, Adhiparasakthi Dental College and
Hospital, Melmaruvathur, Tamil Nadu, India

Md Enamul Hoque
Department of Biomedical Engineering, Military Institute of
Science and Technology, Dhaka, Bangladesh

K.M. Ramkumar
Department of Biotechnology, School of Bioengineering,
SRM Institute of Science and Technology, Kattankulathur,
Tamil Nadu, India

5.1 INTRODUCTION

Cultivating animal cells *in vitro* has proven valuable when studying cell structure and function under controlled conditions. Furthermore, cultured cells are used in chromosome karyotyping, the production of vaccines, and hybridoma production. Any kind of tissue can be cultured as long as it is dispersed appropriately, but unlike normal adult tissues, embryonic and tumor tissues are more commonly successful in cell culture.

Those cultures originating from fresh tissues are referred to as primary cultures. A primary or secondary culture derived from normal tissue can only live a finite number of hours or days, like its *in vivo* counterpart. A few cells in a population of millions survive beyond this definite lifespan and acquire an indefinite capacity to divide; these cells are called cell lines. Several cancer cells are capable of dividing indefinitely in culture. Viruses and chemical carcinogens can also transform normal cells into continuous cell lines [1].

Tissues with a significantly higher proportion of cells are usually used to prepare primary cultures since these tissues contain a larger number of differentiated cells. It is easier to disaggregate cells and produce more viable cells from embryonic tissues,

DOI: 10.1201/9781003397755-6

which are preferred for primary cultures. Primary cultures should contain more cells since their survival rate is much lower.

5.2 MATERIALS

- Tissues
- Calcium, magnesium-free – phosphate-buffered saline (PBS)
- Trypsin (0.25%)
- Trypsinization flask
- Ethylenediaminetetraacetic acid (EDTA)
- Complete medium
- CO_2 incubator
- BP blade
- Forceps, Petri plates, and 100 ml conical flasks

5.3 METHODOLOGY

1. Before the experiments, the tissues were processed in a biosafety cabinet, and proper sterilization and materials required were followed.
2. For 2–3 minutes, the tissues were washed in phosphate-buffered saline solution (pH 7.3–7.4) and rinsed in complete media with antibiotics for possible disinfection.
3. A sterile blade was used to mince the tissues into small pieces of 1 mm × 1 mm size and then placed in a sterilized Petri dish containing complete media.
4. The cells were separated using Trypsin (0.25%) containing 0.02% EDTA.
5. A centrifuge minced the tissues for 3 minutes at 4000 rpm to recover pellet.
6. Fresh medium was added to the pellet, and the resuspended pellet was seeded into six-well culture plates.
7. The culture plate was then incubated at 37°C in a humidified incubator supplemented with 5% CO_2.

5.4 KEY POINTS

1. When transferring primary tissues to cultures, avoid contamination.
2. The incorrect salt content in the culture medium, bacterial contamination, low bicarbonate buffering, or incorrect carbon dioxide tension can cause pH shifts.
3. Changes in the pH of the medium, depletion of essential growth-promoting components/factors, contamination, improper storage of reagents, etc., can cause slow growth.
4. Increased concentration of toxic metabolites, temperature fluctuations, lack of CO_2, and imbalanced osmotic pressure in the culture medium resulted in lowered cell survival [2].
5. Using a frozen medium, residual phosphate remaining after washing with detergent can cause powdered medium components to precipitate in the medium without changing the pH.

6. Calcium, magnesium ions, or released DNA (over-digesting with proteo-lytic enzymes) can cause suspension cells to clump [3].

7. Data obtained using primary cells are subject to variability due to the use of various reagents and media. Variability may also be caused by the handling methods used by users.

REFERENCES

[1] Freshney, R.I. 2005. *Culture of Animal Cells: A Manual of Basic Techniques.* 5th ed. Hoboken, New Jersey. John Wiley & Sons Inc.

[2] Golnar, A. et al. 2018. Basics of Cell Culture: A Student Laboratory Manual. Volume 7. Edited by Deborah Blancas, Carin Zimmerman, Ph.D., and Edie Kaeuper, Ph.D. (2011).

[3] Segeritz, C.P., Vallier, L. et al. 2016. Cell Culture: Growing Cells as Model Systems In Vitro *Basic Science Methods for Clinical Researchers.* 10.1016/B978-0-12-803077-6.00009-6

6 Passaging of Cell Lines

Milan K.L. and Goutham V. Ganesh
Department of Biotechnology, School of Bioengineering,
SRM Institute of Science and Technology, Kattankulathur,
Tamil Nadu, India

Dhamodharan Umapathy
Department of Biotechnology, School of Bioengineering,
SRM Institute of Science and Technology, Kattankulathur,
Tamil Nadu, India

Department of Research, Adhiparasakthi Dental College and
Hospital, Melmaruvathur, Tamil Nadu, India

Md Enamul Hoque
Department of Biomedical Engineering, Military Institute of
Science and Technology, Dhaka, Bangladesh

K.M. Ramkumar
Department of Biotechnology, School of Bioengineering,
SRM Institute of Science and Technology, Kattankulathur,
Tamil Nadu, India

6.1 INTRODUCTION

The process of subculturing, also known as passaging, involves taking out the medium from a previous culture and moving the cells into a fresh culture, a procedure of further propagating the cell line or cell strain. In culture, the cells proliferate exponentially from the lag phase following seeding to the log phase. It is not uncommon for the proliferation of cells to cease when they utilize all the substrate available for growth in adherent culture or when they exceed the capacity of the medium to support further growth in suspension culture. A fresh medium must be supplied to the culture to maintain its optimal density and stimulate further proliferation.

Cells can be grown in culture as monolayers on a substrate (adherent culture) or free-floating in a culture medium (suspension culture). Most cells derived from vertebrates need to be cultured on a substrate specifically treated so that cell adhesion and spread can take place (i.e., tissue-culture treated), except for hematopoietic cell lines and a few others. Suspension culture is, however, an option for many cell lines [1].

Adherent and suspension cultures use similar criteria to decide whether to subculture, though mammalian and insect cell cultures have some differences.

DOI: 10.1201/9781003397755-7

In the log phase, adherent cultures should be passaged before they reach confluence. It takes longer for normal cells to recover after reseeding when they are at the confluence (contact inhibition). Cells in suspension should be passaged when they are in log-phase growth before they reach confluency [2]. When confluency occurs, the medium in the culture flask appears turbid when it is swirled.

In most cases, a drop in pH indicates lactic acid buildup due to cellular metabolism. A decreased pH can be suboptimal for cellular growth since lactic acid is toxic. Cultures at a higher cell concentration generally exhaust medium more rapidly than cells at a lower concentration since the pH changes faster. When the pH drops rapidly (>0.1–0.2 pH units) and cell concentration increases, you should subculture your cells [3].

You can monitor the health of your cells by passaging them on a precise schedule to ensure reproducible behavior. Seed your cultures at varying densities until you achieve consistent growth rates and yields for your cell type. Growth patterns that differ from these are usually indicative of unhealthy cultures (e.g., deterioration, contamination) or a component of the culture system malfunctioning (e.g., poorly controlled temperature, aged culture medium). Keeping a detailed cell culture log is highly recommended. This log should include feeding and subculture schedules, the type of media used, the dissociation procedure used, split ratios, morphological observations, seeding concentrations, yields, and any antibiotics needed. Experiments and other non-routine procedures (i.e., changing the media) should be performed according to the prior chosen schedule. Make sure not to advance your experiments during the lag period or after they have reached confluency and ceased to grow if your experimental schedule does not meet the regular subculture schedule.

6.2 PASSAGING OF ADHERENT CELL LINES

6.2.1 MATERIALS

- Pre-warmed media to 37°C
- Isopropanol 70% (v/v) in sterile water
- Ca^{2+}/Mg^{2+} free PBS
- 0.5% trypsin/EDTA in HBSS (Hank's balanced salt solution)
- Trypan-blue (vital stain)
- Water bath
- Cabinet with an appropriate level of containment for microbiological safety
- CO_2 incubator
- Pre-labeled flask
- Inverted phase-contrast microscope
- Cooling centrifuge
- Haemocytometer
- Pipettes

6.2.2 METHODOLOGY

1. Using PBS, wash the cell monolayer. Repeat if the cells adhere strongly.
2. Pour approximately 1 ml of 0.5% trypsin/EDTA per 25 cm^2 of the surface area onto the cell monolayer. Cover the monolayer with Trypsin and gently agitate the flask.
3. Incubate the flask for 2–10 minutes in an incubator.
4. Make sure all the cells are floating and detached using an inverted microscope. To release any remaining attached cells, gently tap the flask's side.
5. Trypsin must be inactivated by resuspending the cells in fresh medium containing serum. Remove 100–200 µl and perform a cell count.

(Note: To inactivate Trypsin in cells grown in serum-free media, use an inhibitor of Trypsin, for example, soybean trypsin inhibitor.)

6. Incorporate cells into a flask containing a pre-warmed medium.
7. Incubate according to the cell line as required by the cell line's growth characteristics, and repeat this process.

6.2.3 KEY POINTS

1. To prevent premature dissociation of cells when growing as attached lines, the culture medium must be maintained, and flasks handled carefully.
2. When present alone, Trypsin detaches cells, but EDTA increases its activity upon removing inhibitory cations.
3. The trypsin enzyme is inactivated in serum, so it is imperative to remove all traces of serum from the cell monolayer by washing it in PBS without Ca^2+/Mg^2+.
4. EDTA and Trypsin should only be used for a short period to detach cells. Cell surface receptors may be damaged by prolonged exposure.
5. If Trypsin is not neutralized with serum, cells will not attach to flasks.
6. You can also neutralize Trypsin by adding soybean trypsin inhibitor, typically available as 1 mg/ml solution, to an equal volume of trypsinized cells [4]. Once centrifuged, the cells are resuspended in a fresh culture medium and counted. Cell cultures without serum require this in particular.

6.3 PASSAGING OF SUSPENSION CELL LINES

6.3.1 MATERIALS

- Pre-warmed media to 37°C
- Alcohol 70%
- Trypan-blue
- Safety equipment (safety visors, sterile gloves, lab coats,)
- Water bath

- Cabinet with an appropriate level of containment for microbiological safety
- Centrifuge
- Incubator
- Inverted phase-contrast microscope
- Haemocytometer
- Pre-labeled flasks
- Pipettes

6.3.2 METHODOLOGY

1. With an inverted phase-contrast microscope, observe cultures for bright, round, refractile cells in the exponential phase. Hybridomas may be sticky and need to be gently knocked from the flask. In EBV-transformed cells, huge clumps may grow, which can be hard to count.
2. Use a hemocytometer to count a sample of cells from the suspension.
3. Determine the number of cells and reseed them into fresh flasks using diluted cells without centrifugation.
4. Repeat the process every 2–3 days.

6.3.3 KEY POINTS

1. Ensure that the spent media for analysis is retained if the cell line is a hybridoma or similar cell line that produces some substance of interest (e.g., recombinant proteins or growth factors).
2. When cell concentration is necessary, centrifuge at $150 \times g$ for 5 minutes, then resuspend in an appropriate volume and count using a hemocytometer before transferring cells to the culture vessel.

REFERENCES

[1] Freshney, R.I. 2005. *Culture of Animal Cells: A Manual of Basic Techniques.* 5th ed. Hoboken, New Jersey: John Wiley & Sons Inc.
[2] Geraghty, R.J., Capes-Davis, A., Davis, J.M., Downward, J., Freshney, R.I., Knezevic, I., et al. 2014. "Guidelines for the use of cell lines in biomedical research". *British Journal of Cancer.* 10.1038/bjc.2014.166.
[3] Segeritz, C.P., Vallier, L. et al. 2016 Cell Culture: Growing Cells as Model Systems *In Vitro. Basic Science Methods for Clinical Researchers.* 10.1016/B978-0-12-803 077-6.00009-6.
[4] Golnar, A. et al. 2018. Basics of Cell Culture: A Student Laboratory Manual. Volume 7. Edited by Deborah Blancas, Carin Zimmerman, Ph.D., and Edie Kaeuper, Ph.D. (2011).

7 Cell or Tissue Culture Models and Cell Line Immortalization

Goutham V. Ganesh and Milan K.L.
Department of Biotechnology, School of Bioengineering, SRM Institute of Science and Technology, Kattankulathur, Tamil Nadu, India

Dhamodharan Umapathy
Department of Biotechnology, School of Bioengineering, SRM Institute of Science and Technology, Kattankulathur, Tamil Nadu, India

Department of Research, Adhiparasakthi Dental College and Hospital, Melmaruvathur, Tamil Nadu, India

K.M. Ramkumar
Department of Biotechnology, School of Bioengineering, SRM Institute of Science and Technology, Kattankulathur, Tamil Nadu, India

7.1 BASIC CONCEPTS

The casual reference to tissue culture when actually referring to cell culture is very common, but this fundamentally underlines the origins of the latter from the former. Cell culture, in the conventional sense, consists of growing cells in a single cell suspension (either adherent or in suspension) that can be derived either directly from organs or tissues (primary culture) and passaged repeatedly (sub-culture for a few times owing to their finite lifespan) or such cells transformed through viral (transduction) or other means to gain immortality that in turn can be propagated indefinitely. The different terms used here also possibly refer to the continuum [1]: culture of aggregated 3D networks of tissues (organ culture) that break down into single dispersed cells (hence, cell culture) that can also reform a network of cells from the same culture (histotypic culture) or resembling an organ with the different cell of different lineages in the niche (organotypic culture).

The primary culture of cells derived from organs or tissues from animals is a tedious process that usually encounters various problems, including accessing a source that supplies adequate numbers of cells for culture, complete disaggregation

DOI: 10.1201/9781003397755-8

25

of tissue with minimal loss of cellular viability, and, importantly, microbial contamination. Besides, primary cultures usually undergo cell division a finite number of times before reaching senescence. Primary cells can therefore be sub-cultured for a limited number of times. To ensure their propagation for infinite cell divisions, such cells can be transformed to yield *continuous cell lines* by immortalization. Both primary cells and cell lines are used extensively in biomedical research and industrial applications, depending on the suitability for the required application. Most common cell lines include the human cervical cancer cell line HeLa, human embryonic kidney cells HEK-293, mouse 3T3 fibroblasts, and several others. Of course, in addition to problems witnessed with primary cultures such as microbial contamination, cell lines under repeated passaging acquire key changes in their physiological functioning and no longer adequately represent their source [2]. Furthermore, cell lines also require molecular characterization owing to the problem of either cross-contamination with other cell lines or misidentification.

In further continuation of aspects dealing with continuous cell lines, those derived from tumor biopsies from patients, such as HeLa discussed above or the human monocytic cells (THP1) obtained from a patient suffering from acute monocytic leukemia, rose from a spontaneous mutation during carcinogenesis and demonstrate immortality even after several decades since their isolation and initial culture. However, many other cell lines available today were initially primary cells that were transformed virally or other methods to acquire an immortal state and thereby propagate indefinitely. Several signaling pathways mediate the cellular senescence process, such as the Wnt/β catenin, MAPK, PI3K/Akt, and TGF-β pathways, have been reported [3]. The Simian virus 40 (SV40) is commonly used to immortalize many cells by manipulating the host cell machinery governing cellular proliferation, tumor suppression, and cell death in both human and rodent cells. Shay et al. reported the frequency at which SV40 induces immortalization in diploid human cells (10^{-8} to 10^{-5}), which was much lower than in rodent cells (nearly 100%). Of course, the length of telomeres would also need to be maintained (above the hay flick limit) for sustained cell division, and several mechanisms dependent or independent of telomerase activation have been reported [4].

Human fibroblasts are particularly noted for acquiring senescence after several passages to cease cellular proliferation. Tsutsui et al. used non-transformed human fibroblasts (WHE-7 cells) and, in addition, fibroblasts isolated from the skin of a patient suffering from a disease with p53 (regulator of the cell cycle) to confer cell immortalization by irradiation with X-rays [5]. A lentivirus vector was also reported [6] to induce reversible transformation of primary quiescent liver endothelial cells with possible benefits in devising novel therapeutic approaches. This may be particularly important, at least in part, owing to the ECM changes in normal versus immortalized cells that, in turn, would prove consequential in clinical therapies involving the administration of immortalized fibroblasts, for instance, to aged patients suffering from skin tissue ailments that require regeneration. A general methodology [7] for viral vector-induced immortalization of human primary epithelial cells is described below:

1. Thaw the recombinant lentiviral vector (with or without selection markers since lentivirus-mediated cell immortalization often results in the generation of

stable cell lines and non-immortalized primary cells eventually die in culture) in a 37°C water bath rapidly and remove it immediately upon thawing.

2. Meanwhile, add polybrene solution before introducing the viral vector to the primary cells at a final concentration of 10 μg/mL and incubate for 24 h. Please note polybrene solution is toxic to some primary cultures, especially at higher concentrations and with prolonged times of incubation. Therefore, the nontoxic concentration may be determined before using polybrene for immortalization.

3. Optional: About 6–8 hrs post-exposure to the vector, replace with fresh medium containing a second dose of the vector with polybrene until the next day.

4. Replace the viral vector-containing medium with a normal growth medium the following day and incubate cells under standard cell culture conditions.

5. About 3–4 days later, sub-culture cells onto sterile plastic dishes with or without appropriate selection drug (transformed cells containing the viral vector incorporating the select antibiotic resistance gene will survive in antibiotic culture medium, but since lentiviral transformed cells in general tend to stably integrate the virus in their genomes, antibiotic selection may not be required).

6. Allow cells to grow for 10–15 days and validate cultures for successful transformation via assessing for viral vector (transgene) expression via PCR or other techniques.

REFERENCES

[1] Ian, Freshney 2005. *Culture of Animal Cells. A Manual of Basic Technique and Specialized Applications.* 6th edition. WILEY, UK.

[2] Hughes et al. 2007. The costs of using unauthenticated, over-passaged cell lines: how much more data do we need? 10.2144/000112598

[3] Wang, Y., Chen, S., Yan, Z. et al. 2019. A prospect of cell immortalization combined with matrix microenvironmental optimization strategy for tissue engineering and regeneration. *Cell Bioscience* 9, 7. 10.1186/s13578-018-0264-9

[4] Shay, J.W., & Wright, W.E. 1874. Quantitation of the frequency of immortalization of normal diploid fibroblasts by SV40 large T-antigen. *Experimental Cell Research*, 1989, 109–118.

[5] Tsutsui, T., Tanaka, Y., Matsudo, Y., Hasegawa, K., Fujino, T., Kodama, S., & Barrett, J.C. 1997. Extended lifespan and immortalization of human fibroblasts induced by X-ray irradiation. *Molecular Carcinogenesis*, 18(1), 7–18. 10.1002/(sici) 1098-2744(1997)

[6] Salmon, P., Oberholzer, J., Occhiodoro, T., Morel, P., Lou, J., & Trono, D. (2000). Reversible immortalization of human primary cells by lentivector-mediated transfer of specific genes. *Molecular Therapy: The Journal of the American Society of Gene Therapy*, 2(4), 404–414. 10.1006/mthe.2000.0141

[7] General Guidelines for Cell Immortalization. *Cell Immortalization Handbook.* Volume 7, Applied Biological Materials Inc.

8 Counting of Cells by Hemocytometer

Milan K.L., Goutham V. Ganesh,
Sundhar Mohandas, and K.M. Ramkumar
Department of Biotechnology, School of Bioengineering,
SRM Institute of Science and Technology, Kattankulathur,
Tamil Nadu, India

8.1 INTRODUCTION

Developed by Louis-Charles Malassez (), the hemocytometer is used to count cells. When viewed from a distance, the hemocytometer resembles a glass slide, except that it is much more than that. A special coverslip is also included, which is unique. Marked grooves appear on the slide in the form of an H with the horizontal line separating the two grids for counting. Therefore, each slide has two identical grids for counting cells. Both grids are 0.1 mm deep. Each grid has a measurement of 3 × 3 mm^2, a square with three equidistant horizontal and vertical lines. These smaller squares, equal to 1 mm^2 each, are then divided into nine smaller squares. A triple-ruled line separates them. Counting bigger cells is done with the four corner squares, while counting smaller cells is done with the center square [1-2].

There are 16 smaller cells within the four corners of the main grid. Within the main grid, the central square is divided into 25 smaller squares, each further divided into 16 smaller squares. After the coverslip has been placed, the sample to be counted is loaded onto the slide. The grooves on each side allow excess fluid to drain. However, when loading a sample, the person must take extreme care. A coverslip covers the grid, creating a chamber. There is a 0.1 ml capacity per larger square in the grid (Figure 8.1) [3].

The cells must be trypsinized first before counting using a hemocytometer. A calculation of cell concentration can be achieved as the volume of each square is known.

8.2 MATERIALS

- Pipettes
- Hemocytometer
- Trypsin; warmed
- Sterile PBS; warmed
- Complete media; warmed

DOI: 10.1201/9781003397755-9

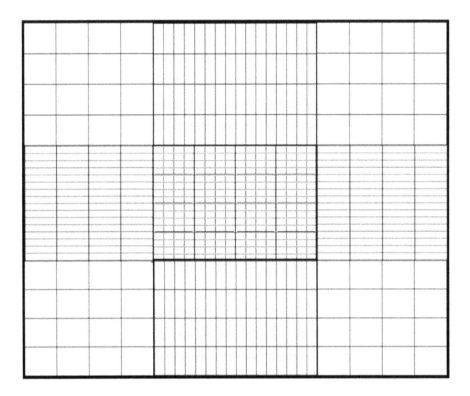

FIGURE 8.1 Schematic representation of the grids of a hemocytometer.

8.3 METHODOLOGY

The following steps should be taken when counting cells using a hemocytometer:

1. Wipe down the hood with alcohol after turning it on.
2. Remove the media from the flask, and rinse the cells once with 3 ml of PBS. You should remove the PBS.
3. To detach cells, add 1 ml trypsin and let stand for 2–3 minutes. To loosen the cells, tap the flask on the side. Examine the cells under a microscope to ensure that they are all floating. To quench trypsin, add 4 ml of media. Using a pipette, mix well.
4. Fill each chamber with 10–12 μL of the cell suspension using a micropipette. Placing the pipette tip on the edge of the coverslip, hold the coverslip loosely with your thumb and release the liquid. Follow the same procedure for the other half.
5. A hemocytometer should be placed on a microscope stage. Locate the grids with a low magnification (4X). To view each square and cell clearly, change the magnification to a higher level (usually 10X will work best).
6. There are five squares in the grid: four corner squares and one in the middle. Count the number of cells within those five squares. Repeat

counting five squares for the other grid. Consider the cells lying on the outside of both border lines of a given square, and do not consider the cells lying inside the border lines. Calculate the number of cells in the center square by counting the cells that lie on any two borders, excluding the cells on the other two borders. Be consistent in counting and consider which borders are to be counted and which are to be excluded.

7. A proper count should consist of approximately 10–50 cells/square. If you have too many cells, the chances of making mistakes are high. Dilute your sample and repeat counting. To make your counts statistically valid, you must spin your cells and then resuspend your pellet in less media volume before counting again.

8. Based on the following formula, calculate cell concentration (cells/ml):

$$\frac{(\text{Total \# cells counted})(10)^4(\text{dilution factor})}{(\text{Total \# squares counted})}$$

8.4   KEY POINTS

1. In addition to being fragile, hemocytometers are also expensive. Make sure you don't drop and break them.

2. Cells in the flask or tube tend to settle to the bottom quickly, and they tend to bond together and form clumps. Counts are distorted by clumps. To guarantee a homogeneous cell suspension, it is very important to pipette up and down before counting the cells to dissociate clumps.

3. If the cells are not plated quickly after being removed from the incubator and trypsinized, they will die from a stress reaction. Thus, it is sometimes more important to speed up a process rather than ensure it is accurate. For those who are short of time or wish to speed up the process, you may choose to count cells in five rather than ten squares.

4. Check the cultured cells before you trypsinize for counting. Counting may not be accurate if confluence is low. Then, after the cells are trypsinized and quenched, a few can be taken out in a tube for counting. After that, quench the remaining cells in the flask with media. In the tube, the cells tend to be more concentrated, so you have a better chance of getting a large number for your count. If the concentration of your sample is too high, it can always be diluted. The cells can also be concentrated after quenching trypsin if you have too few cells. You can resuspend your cells in less volume of media by transferring them to a centrifuge tube, spinning them for 3 minutes, and then removing the supernatant. Then count again.

5. Depending on how confluent the flask is, you might need to quench trypsin with an additional volume of media. Your culture determines how much media to consume. It is necessary to refrain from adding too much media as too much dilution will make the count difficult. Keep in mind that you are always better off diluting than concentrating.

6. Adding a vital dye called Trypan Blue to the cells before counting allows you to distinguish between dead and live cells. Add 50 μl of Trypan Blue to 50 μl of cells and pipette the mixture into the hemocytometer.

7. The integrity of the cell membrane of dead cells is compromised; therefore, Trypan Blue stains dead cells blue. Healthy cells do not take up the dye and hence do not stain with Trypan Blue. The cells must be counted immediately after mixing the Trypan Blue with them because live cells will eventually die and turn blue in the solution.

8. Take into account the dilution factor of 2, since the cells were diluted with 2X Trypan Blue solution.

REFERENCES

[1] ML Verso Some Nineteenth-Century Pioneers of Haematology. 1971.; 15(1): Medical History, 55–67.

[2] Golnar Afshar. Basics of cell culture. A student laboratory manual. Volume 7. Edited by Deborah Balances, Carin Zimmerman. (2011).

[3] Absher, Marlene 1973. Hemocytometer counting, 395–397, Academic Press. 10.1016/B978-0-12-427150-0.50098-X

9 Cryopreservation of Cells

Milan K.L., Karan Naresh Amin, and Sundhar Mohandas

Department of Biotechnology, School of Bioengineering, SRM Institute of Science and Technology, Kattankulathur, Tamil Nadu, India

Md Enamul Hoque

Department of Biomedical Engineering, Military Institute of Science and Technology, Dhaka, Bangladesh

K.M. Ramkumar

Department of Biotechnology, School of Bioengineering, SRM Institute of Science and Technology, Kattankulathur, Tamil Nadu, India

9.1 INTRODUCTION

It is common for continuous cell lines to drift genetically. Finite cultures are subjected to senescence. All cell cultures are likely to be contaminated by microbes, and even the most tightly managed laboratories fail when their equipment malfunctions. Recognizing the value of established cell lines and the difficulty of replacing them, they must be preserved frozen for long-term storage.

In the event of an excess of cells after sub-culturing, they should be preserved as seed stock, kept protected, and not used in general laboratory work. From these preserved seed stocks, working stocks may be formed and replenished as needed. Cryopreserved working stock can be used to prepare fresh seed stock when the seed stocks are depleted.

It is imperative to check and identify for contamination of mammalian cells before cryopreservation to avoid contamination loss and minimize genetic changes. Continuous cell lines should also be cryopreserved to minimize the risk of aging and transformation. Freezing cells can be accomplished using many different media.

For cryopreserved cells, dimethylsulfoxide (DMSO) is the best agent used in addition to liquid nitrogen. As a result of cryoprotective agents, the freezing point of the medium is lowered, and the cooling rate is slowed, which helps reduce the formation of ice crystals capable of damaging cells and causing cell death [1].

Note: It is known that organic molecules can penetrate tissues using DMSO as a solvent. Use appropriate equipment when handling and disposing of reagents

DOI: 10.1201/9781003397755-10

containing DMSO. Ensure that all waste streams are properly disposed of according to local regulations.

9.2 GUIDELINES FOR CRYOPRESERVATION

As with other procedures, you will need to follow the instructions provided with your cell lines to preserve your cells for future use.

The following are the guidelines for cryopreserving cell lines:

- Cells cultured at low passage number and high concentrations should be frozen. Before freezing, ensure that cell viability is at least 90%. It is important to note that the best freezing conditions vary with the cell line.
- Use a controlled-rate cryofreezer or a cryofreezing container to gradually freeze the cells at a temperature of about 1°C per minute.
- Ensure you are using the recommended freezing medium. DMSO or other cryo-protective agents should be present in the freezing medium.
- The freezing temperature should be less than −70°C as frozen cells will degrade above −50°C.
- When storing frozen cells, make sure you use sterile cryovials. Cells are frozen in cryovials and either submerged in liquid nitrogen or stored in a gaseous phase above the liquid nitrogen.
- Make sure you wear personal protective equipment (PPE).
- Sterilization must be adhered to whenever solutions and equipment come into contact with cells. Use Laminar flow hoods and proper sterilization techniques [2].

9.2.1 SERUM-CONTAINING MEDIUM CONSTITUENTS

- 10% glycerol in the complete medium or 10% DMSO (dimethyl sulfoxide) in the complete medium.
- The medium will be a 50% cell-conditioned medium and 50% fresh medium containing 10% glycerol or 10% DMSO.

In general, cryopreservatives and protein sources are present in the cryopreservation media. Cells are protected from freezing and thawing by cryopreservatives and proteins. Serum-free medium, although generally lacking in protein, can still be used as a base for a cryopreservative medium in the following formulations [3].

9.2.2 SOME OF THE COMMON MEDIA CONSTITUENTS FOR SERUM-FREE MEDIUM

- 50% cell-conditioned serum-free medium with 50% fresh serum-free medium containing 7.5% DMSO or the medium should consist of 7.5% DMSO and 10% cell culture grade BSA.

9.3 MATERIALS

- The vessels should contain cultured cells during the log phase of cell growth.
- Complete growth medium
- DMSO or a freeze-preserving medium available commercially (open only in a laminar flow hood) are suitable cryoprotective agents
- Conical tubes, 15 or 50 mL (disposable and sterile)
- Reagents and equipment for determining viable and total cell counts (e.g., hemocytometer)
- The cryogenic storage vials (i.e., cryovials) must be sterile.
- Isopropanol chamber or controlled-rate freezing apparatus
- Liquid nitrogen storage container

Aside from the materials listed above, for freezing adherent cells, you need the following:

1. A balanced salt solution without calcium, magnesium, or phenol red, such as Dulbecco's Phosphate-Buffered Saline (DPBS)
2. Trypsin or TrypLE^{TM} Express enzyme, without phenol red as dissociation reagent

9.4 METHODOLOGY

The following methodology is a general procedure for cryopreserving cultured cells.

1. Prepare a freezing medium and store it in the refrigerator at 2–8°C until used. Note: You must consider the type of cell line when selecting the freezing medium.
2. Remove adherent cells from the tissue culture vessel by gently separating them from the vessel with the same technique used when sub-culturing. Resuspend cells in the medium according to the requirement of the cell type
3. Using a hemocytometer, determine the number of cells and percent viability. Estimate the volume of freezing medium needed according to the desired viable cell density
4. The cell suspension is centrifuged at approximately 100–200 × g for 5–10 minutes. Ensure that the cell pellet is not disturbed by decanting the supernatant.

   Note: Different types of cells require different centrifugation speeds and durations.

5. In a cold freezing medium, resuspend the cell pellet in the recommended viable cell density for the specific cell type.

6. Fill cryogenic storage vials with aliquots of the cell suspension. When aliquoting cells, mix them frequently to ensure that the suspension remains homogeneous.
7. Decrease the temperature by approximately 1°C per minute while freezing the cells in a controlled-rate freezing b apparatus. Alternatively, in an isopropanol chamber, place the cryovials containing the cells overnight at −80°C.
8. Transfer frozen cells to liquid nitrogen, and store them in the gas phase above the liquid nitrogen.

9.4.1 METHODOLOGY: SUSPENSION CULTURES

1. Estimate the number of viable cells that should be cryopreserved. It is recommended that the cells be in log phase. To pellet cells, centrifuge them at 200–400 × g for 5 minutes. Pipette the supernatant to the smallest volume possible without disturbing the cells.
2. Cells should be resuspended in freezing medium at a concentration of 2×10^7 to 5×10^7 per mL when using serum-containing medium, or $0.5 \times 10^7 \times 10^7$ to $1 \times 10^7 \times 10^7$ per mL when using serum-free medium.
3. Aliquot into cryogenic storage vials. Place vials on wet ice or in a 4°C refrigerator, and start the freezing procedure within 5 minutes.
4. Cells are frozen slowly at 1°C/min. This can be achieved using a programmable cooler or by placing vials in an insulated box, a −70°C to −90°C freezer, then transferring them to liquid nitrogen storage.

9.4.2 METHODOLOGY: ADHERENT CULTURES

1. Detach cells from the substrate with dissociation agents. Detach as gently as possible to minimize damage to the cells.
2. Resuspend the detached cells in a complete growth medium and establish the viable cell count.
3. Centrifuge at ~200 × g for 5 minutes to pellet cells. Using a pipette, withdraw the supernatant to the smallest volume without disturbing the cells.
4. Resuspend cells in a freezing medium to a concentration of 5×10^7 to 1×10^8 cells/mL.
5. Aliquot into cryogenic storage vials. Place vials on wet ice or in a 4°C refrigerator, and start the freezing procedure within 5 minutes.
6. Cells should be frozen slowly at 1°C/min. This can be achieved using a programmable cooler or by placing vials in an insulated box placed in a −70°C to −90°C freezer, then transferring them to liquid nitrogen storage.

REFERENCES

[1] Hunt Charles J., Crook, Jeremy M., Ludwig, Tenneille E. (2017). Cryopreservation: Vitrification and Controlled Rate Cooling, Stem Cell Banking: Concepts and Protocols, Methods in Molecular Biology, New York, Springer, vol. 1590, pp. 41–77, 10.1007/978-1-4939-6921-0_5

[2] Freshney, R. (1987). *Culture of Animal Cells: A Manual of Basic Technique*, p. 220, Alan R. Liss, Inc.: New York.

[3] Jang, Tae Hoon, Park, Sung Choel, Yang, Ji Hyun, Kim, Jung Yoon, Seok, Jae Hong, Park, Ui Seo, Choi, Chang Won, Lee, Sung Ryul, & Han, Jin (2017). Cryopreservation and its clinical applications. *Integrative Medicine Research*, 6, 12–18 10.1016/j.imr.2016.12.001.

10 Thawing of Cells

Milan K.L., Goutham V. Ganesh,
Sundhar Mohandas, and K.M. Ramkumar
Department of Biotechnology, School of Bioengineering,
SRM Institute of Science and Technology, Kattankulathur,
Tamil Nadu, India

10.1 INTRODUCTION

Cells should be warmed from a frozen state as quickly as possible until they are completely thawed. Placing the frozen vial in a 37°C water bath will allow the vial to warm rapidly. Remember that content frozen in plastic vials will take longer to thaw than material frozen in glass ampoules; you may be able to accelerate the melting process by gently agitating the vial while it is being heated. When handling cells that are susceptible to physical or mechanical stress, adequate care should be taken to ensure minimal agitation. The vial must be removed from the water bath as soon as the contents have been thawed. Before opening the vial, wipe its external surface with alcohol-soaked gauze to minimize the risk of contamination during reconstitution. To minimize exposure to the cryoprotective agent, the contents of the vial should be transferred immediately into fresh growth media following thawing. If using cell lines, you may want to dilute the contents of the vial before using it, but for most cultures, it can be used directly in fresh media. After initial dilution, the cell suspension is centrifuged at $100 \times g$ for 10 minutes, the supernatant is removed, and the cells are resuspended in fresh growth media to eliminate any remaining cryoprotectants [1]. Thawing and refreezing may be acceptable in some cases for materials that are not sensitive to cryopreservation. Unless they are in a form that can withstand freezing, most replicable cells will not tolerate refreezing.

10.2 GUIDELINES FOR THAWING CELLS

A frozen culture undergoes a stressful thawing procedure. It is important to work quickly and use good techniques to maximize the survival of the cells. The following methodology yields good results in cell revival.

- Rapidly thaw down the frozen cryovial in warm water in less than a minute.
- You can incubate thawed cells in a pre-warmed growth medium by slowly diluting them before incubation.
- Optimize recovery of thawed cells by plating them at high density.

DOI: 10.1201/9781003397755-11

- Utilize laminar flow hoods and use proper aseptic procedures.
- Make sure you wear protective equipment such as goggles or a face mask. The thawing of cryovials in the liquid phase can cause explosions [2].
- It has been found that DMSO can enhance the entry of organic molecules into tissues when used in freezing media. It is highly recommended to handle DMSO-containing reagents using appropriate equipment [1].

10.3   MATERIALS

- Cells frozen in a cryovial.
- A 37°C pre-warmed complete growth medium
- Centrifuge tubes, disposable and sterile
- Water bath
- 70% Ethanol
- Tissue culture-treated flasks, dishes, or plates

10.4   METHODOLOGY

The general protocol for thawing frozen vials is as follows; however, adhere to the cell-specific instructions provided for a more detailed methodology.

1. Contamination-free water to be set to 37°C or near temperature.
2. Place the cryovial containing frozen cells in water and gently swirl for less than a minute inside the biosafety cabinet.
3. Add the appropriate complete medium five times the frozen cell volume.
4. Centrifuge at 1500 rpm for 5–10 minutes and discard the supernatant.
5. Optionally count the cells on a hemocytometer by adding pre-warm medium to the cell pellet.
6. Transfer the required density of cells in the number of flasks needed with a complete medium.

REFERENCES

[1] M Uhrig Ezquer F, Ezquer M. Improving Cell Recovery: Freezing and Thawing Optimization of Induced Pluripotent Stem Cells. *Cells* 2022. 10.3390/cells11050799
[2] Freshney, R. (1987) *Culture of Animal Cells: A Manual of Basic Technique*, p. 220, Alan R. Liss, Inc.: New York.

11 Detection of Bacterial and Fungal Contamination

Milan K.L. and Karan Naresh Amin
Department of Biotechnology, School of Bioengineering, SRM Institute of Science and Technology, Kattankulathur, Tamil Nadu, India

Dhamodharan Umapathy
Department of Biotechnology, School of Bioengineering, SRM Institute of Science and Technology, Kattankulathur, Tamil Nadu, India

Department of Research, Adhiparasakthi Dental College and Hospital, Melmaruvathur, Tamil Nadu, India

K.M. Ramkumar
Department of Biotechnology, School of Bioengineering, SRM Institute of Science and Technology, Kattankulathur, Tamil Nadu, India

11.1 INTRODUCTION

An infusion of brain hearts and trypticase soy agar with sheep blood is used to cultivate nutritionally fastidious bacteria of clinical origin present in primary tissue cultures or material contaminated by human skin flora and insufficient aseptic technique. A fluid thioglycollate environment supports the growth of micro-aerophilic or slightly anaerobic bacteria; these contaminants are often spore-forming contaminants that originate from inadequately sterilized materials. Ideally, soybean/casein digest broth should be used to grow a wide range of bacteria from humans or the environment. In addition to being a general bacterial growth medium, HEPES/trypticase/yeast extract (HTYE) broth has the advantage of ensuring the growth of nutritionally or physiologically stressed bacteria that would not be easy to cultivate in other media [1]. The bacteria in these general categories arise from the environment and are commonly found in distilled water, filtrate in deionization systems, and airborne contaminants. The Emmons modification is used to detect fungi (mold) in Sabouraud's agar, as is YM agar for yeast detection. The growth of

DOI: 10.1201/9781003397755-12

mold is frequently impacted by environmental factors. A constant rain of micro-spores from the spores often occurs in laboratories because they live in air-handling ducts [2]. It is common to find mold growing on the tubing dispensing distilled water from reservoirs. In most cases, yeasts associated with cell culture contami-nants are of human origin.

11.2   MATERIALS

- A broth with an aerobic nutrient mix, e.g., soyabean casein digest (Tryptone Broth, TSB) (15 ml aliquots)
- Fluid Thioglycollate Medium (20 ml aliquots) (TGM): anaerobic nutrient broth
- Positive control organisms
- Water bath (37°C)
- Incubator

11.3   METHODOLOGY

1. Before testing, culture the cell line without antibiotics for two passages.
2. Use a cell scraper to bring attached cells into suspension. Suspension cell lines may be tested directly.
3. Using 1.5 ml of the test sample, inoculate 2 × aerobic and 2 × anaerobic broths.
4. You should inoculate 2 × aerobic broths and 2 × anaerobic broths with each positive control organism (e.g., 0.1 ml control at 100 CFU per broth).
5. With 1.5 ml of sterile PBS, inoculate 2 × aerobic and 2 × anaerobic broths
6. Incubate broths as follows:
   Incubate TGM broths at 32°C and TSB broths at 22°C
7. Examine the test broths on days:
   - 3, 4, or 5
   - 7 or 8
   - 14
   - Control broths are read at 7 days

Criteria for a Valid Result: After 7 days of incubation, bacteria and fungi are detected in all positive control broths, while the negative and PBS control broths do not show any signs of infection.

Criteria for a Positive Result: Bacteria or fungi in test broths display turbidity.

Criteria for a Negative Result: A clear broth should not show any signs of turbidity in the test.

11.4   KEY POINT

A microbiology laboratory should be used in place of a cell culture lab for this test procedure.

REFERENCES

[1] Freshney, R.I. 1994. Contamination. In *Culture of Animal Cells: A Manual of Basic Technique*, 3rd ed., pp. 243–252. Wiley-Liss: New York.

[2] Hay, R.J., Caputo, J., and Macy, M.L. 1992. *ATCC Quality Control Methods for Cell Lines*, 2nd ed. American Type Culture Collection: Rockville, MD.

12 Testing for Mycoplasma Contamination

Milan K.L. and Ravichandran Jayasuriya
Department of Biotechnology, School of Bioengineering, SRM Institute of Science and Technology, Kattankulathur, Tamil Nadu, India

Dhamodharan Umapathy
Department of Biotechnology, School of Bioengineering, SRM Institute of Science and Technology, Kattankulathur, Tamil Nadu, India

Department of Research, Adhiparasakthi Dental College and Hospital, Melmaruvathur, Tamil Nadu, India

K.M. Ramkumar
Department of Biotechnology, School of Bioengineering, SRM Institute of Science and Technology, Kattankulathur, Tamil Nadu, India

12.1 INTRODUCTION

Aseptic techniques and quality control procedures are important for preventing Mycoplasma infection of cell cultures. Most cultures in a laboratory are infected, typically with the same Mycoplasma species. Infection with Mycoplasma alters cell metabolism significantly [1]. Two cell cultures may react differently to the same kind of Mycoplasma. A single cell culture may be affected differently by two different species of Mycoplasma. Growth rates, amino acids and nucleic acid metabolism, viral growth, interferon levels, karyotype, and other characteristics of cells have been observed [2].

12.2 MATERIALS

- Alcohol 70% (v/v) in sterile water
- Plates of Mycoplasma (in 5 cm Petri dishes)
- Mycoplasma horse serum broths
- *M. orale*
- *M. pneumoniae*
- Water bath (37°C)
- Incubator

DOI: 10.1201/9781003397755-13

12.3 METHODOLOGY

1. Inoculate 0.2 ml of the test sample onto two agar plates.
2. Inoculate 100 CFU of each control organism onto two plates.
3. As a negative control, leave two plates uninoculated.
4. Inoculate 10 ml of the sample into one broth.
5. Inoculate 100 CFU of each control organism into a broth.
6. As a negative control, leave one broth un-inoculated.
7. Incubate agar plates for 2 weeks anaerobically at 36°C.
8. Incubate broths at 36°C aerobically for 14 days.
9. Subculture 0.1 ml of the test broth into the agar plate every 3–7 days and every 10–14 days of incubation, and incubate the plate anaerobically as above.
10. Using an inverted microscope, examine the agar plates after 14 days to determine if Mycoplasma colonies have developed.

Criteria for a Valid Result: On agar plates and broths, positive control agars and broths show typical colonial growth and a color change. Mycoplasma was not detected on any negative control agar plates or broths.

Criteria for a Positive Result: Typical colony formation can be observed on Mycoplasma-infected test agar plates.

Criteria for a Negative Result: Mycoplasma was not detected on the test agar plates.

12.4 KEY POINTS

1. One characteristic colony formation of Mycoplasma is the "fried egg" colony since the central portion of the colony penetrates the agar core surrounded by a flat translucent peripheral zone [3].
2. Biological safety cabinets operating to ACDP Category 2 conditions are required to handle *Mycoplasma pneumoniae*, an important pathogen.
3. The microbiology laboratory should conduct the testing, not the cell culture laboratory.
4. According to the ECACC, detecting Mycoplasma with a minimum of two methods (e.g., indirect DNA staining and culture isolation) is the best way to ensure a more accurate finding because various Mycoplasma species require different detection sensitivity levels.
5. ECACC also offers a PCR assay to detect Mycoplasma strains that will provide a result within 24 hours.

REFERENCES

[1] Stanbridge, E. (1971). Mycoplasmas and cell cultures. *Bacteriological Reviews*, *35*(2), 206–227. 10.1128/br.35.2.206-227.1971
[2] Clyde, W.A., Jr (1964). Mycoplasma species identification based upon growth inhibition by specific antisera. *Journal of Immunology (Baltimore, Md.: 1950)*, *92*, 958–965.
[3] McCarty, G.J. (1975). Detection of mycoplasma in cell cultures. *TCA Manual*, *1*, 113–116. 10.1007/BF01352628

13 Cellular Cytotoxicity

Milan K.L. and Kannan Harithpriya
Department of Biotechnology, School of Bioengineering,
SRM Institute of Science and Technology, Kattankulathur,
Tamil Nadu, India

Dhamodharan Umapathy
Department of Biotechnology, School of Bioengineering,
SRM Institute of Science and Technology, Kattankulathur,
Tamil Nadu, India

Department of Research, Adhiparasakthi Dental College and
Hospital, Melmaruvathur, Tamil Nadu, India

Md Enamul Hoque
Department of Biomedical Engineering, Military Institute of
Science and Technology, Dhaka, Bangladesh

K.M. Ramkumar
Department of Biotechnology, School of Bioengineering,
SRM Institute of Science and Technology, Kattankulathur,
Tamil Nadu, India

13.1 MTT ASSAY

13.1.1 Introduction

Cell-based cytotoxicity assays are widely used in drug development to measure the effect of a drug on the growth of the cell population. Various assay methods are used to estimate the number of viable eukaryotic cells, including tetrazolium reduction, resazurin reduction, protease markers, and ATP detection. Among these, the tetrazolium reduction assay, which is positively charged, readily penetrates viable eukaryotes and measures the enzyme activity of the viable cells. The MTT (3-(4,5-dimethylthiazol-2yl)-2,5-diphenyltetrazolium bromide) is said to be the first homogenous assay suitable for high throughput screening. Firstly, the MTT substrate is prepared and added to the cell at a final concentration of 0.2–0.5 mg/ml and incubated. MTT reagent is reduced in the mitochondria of active cells by the enzyme succinate dehydrogenase to yield water-insoluble formazan crystals. The insoluble formazan crystals precipitate inside the cells and near the cell surface before absorbance reading. The most widely used solubilizing agent in MTT assay

DOI: 10.1201/9781003397755-14

is DMSO. The quantity of formazan is directly proportional to the number of viable cells, measured at 570 nm using a plate reader. The cell that loses its viability fails to form an insoluble purple color; formazan crystal serves as a convenient method to distinguish between viable and non-viable cells [1].

13.1.2 MATERIALS

- MTT reagent
- DMSO
- 96 well plate
- Culture medium
- Basic lab wares

13.1.3 METHODOLOGY

1. Seed trypsinized cells to 96 well plates with the seeding density of 10,000 to 20,000 cells per well.
2. After the end of treatment, remove the treatment medium and wash with sterile PBS.
3. Add 0.5–5 mg/ml of fresh sterile filtered MTT solution to the well and incubate for 4 hours at standard cell culture conditions.
4. After the end of incubation, remove the MTT solution entirely and add the appropriate volume of stock DMSO per well to observe the color change.
5. Measure absorbance between 490 and 570 nm.
6. The percent (%) of viable cells was calculated using the following formula:

$$\% \text{ of viability} = (100 \times (\text{control} - \text{sample}))$$

13.1.4 KEY POINTS

1. Always set up background controls with media and MTT solutions to avoid background reading.
2. Always do the assay in duplicates or triplicates to avoid handling errors.
3. Care should be taken while using the MTT solution. Avoid touching the walls of the plate.
4. Always wrap the plate during the incubation period.

13.2 ALAMAR BLUE ASSAY

13.2.1 INTRODUCTION

Alamar Blue contains the cell-permeable, nontoxic, and weakly fluorescent blue dye resazurin as an indicator dye. Since 1993, this reagent has been widely used and trusted. Compared to the commonly used MTT cell viability assay, Almar Blue is a healthier and nontoxic alternative.

Proliferation can be measured quantitatively with Almar Blue in human, animal, bacterial, fungal, and mycobacterial cells. Cell growth monitoring and cytotoxicity assays can also be conducted on the system, and it is helpful for cytokine bioassays and cell viability tests.

As an electrochemical indicator of oxidation-reduction (REDOX), Resazurin changes color when cellular metabolism is reduced. Resorufin is highly fluorescent in its reduced form, and the amount of fluorescence produced correlates with how many living cells are respiring. Alamar Blue monitors oxidation levels within cells during respiration, assessing the level of cytotoxicity and viability [2].

13.2.2 MATERIALS

1. Alamar Blue
2. PBS
3. DMSO
4. 96 well plates
5. Culture medium
6. Basic lab wares

13.2.3 METHODOLOGY

1. Seed trypsinized cells to 96 well plates with the seeding density of 10,000 to 20,000 cells per well.
2. After the end of treatment, add 4 µl of 0.1% Alamar Blue in PBS and incubate for 4 hours at standard cell culture conditions.
3. After incubation, remove Alamar Blue entirely and add the appropriate volume of stock DMSO per well to observe the color change.
4. Measure absorbance between 570 and 600 nm.
5. The percent of viable cells was calculated using the below-mentioned formula:

$$\text{Percentage of viability} = [(O2 \times A1) - (O1 \times A2) \div (R1 \times N2) - (R2 \times N1)] \times 100$$

where

$O1$ = molar extinction coefficient of oxidized Alamar Blue (blue) at 570 nm
$O2$ = molar extinction coefficient of oxidized Alamar Blue at 600 nm
$R1$ = molar extinction coefficient of reduced Alamar Blue (red) at 570 nm
$R2$ = molar extinction coefficient of reduced Alamar Blue at 600 nm
$A1$ = absorbance of test wells at 570 nm
$A2$ = absorbance of test wells at 600 nm
$N1$ = absorbance of negative control well (media plus Alamar Blue but no cells) at 570 nm

N2 = absorbance of negative control well (media plus Alamar Blue but no cells) at 600 nm

13.2.4 KEY POINTS

1. To avoid fluorescence value errors, adjust the instrument's gain setting and verify the instrument's filter/wavelength setting.
2. Short incubation time and the low number of cells can also cause fluorescence value errors to avoid following the proper incubation period.
3. Bacterial contamination can cause a high fluorescence value. Identify and remove the source of contamination.
4. Care should be taken while using Alamar Blue. Avoid touching the walls of the plate.
5. Always wrap the plate during the incubation period.

REFERENCES

[1] Riss, TL, Moravec, RA, Niles, AL, Duellman, S, Benink, HA, Worzella, TJ, & Minor, L (2004). Cell Viability Assays. Cell Viability Assays.
[2] Rampersad, S.N. Multiple applications of Alamar Blue as an indicator of metabolic function and cellular health in cell viability bioassays. *Sensors (Basel).* 2012;12(9): 12347–12360. 10.3390/s120912347

Section 2

Isolation and Analysis of Nucleic Acids and Proteins

14 Isolation of Total RNA and Assessment of Purity and Concentration

Milan K.L. and Kannan Harithpriya
Department of Biotechnology, School of Bioengineering, SRM Institute of Science and Technology, Kattankulathur, Tamil Nadu, India

Dhamodharan Umapathy
Department of Biotechnology, School of Bioengineering, SRM Institute of Science and Technology, Kattankulathur, Tamil Nadu, India

Department of Research, Adhiparasakthi Dental College and Hospital, Melmaruvathur, Tamil Nadu, India

R. Senthilkumar
Department of Biotechnology, School of Applied Sciences, REVA University, Bangalore, Karnataka, India

K.M. Ramkumar
Department of Biotechnology, School of Bioengineering, SRM Institute of Science and Technology, Kattankulathur, Tamil Nadu, India

14.1 INTRODUCTION

RNA (ribonucleic acid) is made of adenine, guanine, cytosine, and uracil, four nitrogen bases bonded to ribose sugar element, polymeric substance found in many living cells and viruses. All living cells synthesize protein with RNA, and many viruses make use of it to carry their genes. With TRIzol™ Reagent, RNA is isolated and separated from DNA and protein [1]. To facilitate the isolation of small and large RNA molecules, TRIzol™ Reagent contains phenol, guanidine isothiocyanate. As the TRIzol™ Reagent disrupts cells and dissolves their components, it effectively inhibits RNase activity while preserving the integrity of RNA. With TRIzol™ Reagent, RNA, DNA, and proteins can be precipitated sequentially from a single sample. The sample is homogenized using TRIzol™ Reagent, followed by the addition of chloroform, and then allowed to separate into an upper aqueous layer (containing RNA), an interphase,

DOI: 10.1201/9781003397755-16

and a lower organic layer (containing DNA and proteins) [2]. When isopropanol is poured over the aqueous layer, RNA precipitates. Ethanol precipitates DNA from the interphase/organic layer. Isopropanol precipitates the protein from the phenol-ethanol supernatant. RNA, DNA, or a protein that has been precipitated is cleaned to remove impurities, and then resuspended for subsequent use. The RNA extracted with TRIzol™ Reagent is free of protein and DNA contamination.

14.2   MATERIALS

- Water bath or heat block
- Micro centrifuge tubes (1.5 ml)
- Micro pipettes (1000 μl, 200 μl)
- Micro pipette tips (1000 μl, 200 μl)
- TRIzol™ Reagent
- Isopropanol, 100%
- Ethanol, 70%
- RNase-free water
- Human PBMC cells isolated using density gradient centrifugation

14.3   METHODOLOGY

1. For every 0.25 mL of sample volume, add 0.5 ml of TRIzol™ Reagent and homogenize the sample by vortexing.
2. To completely dissociate the nucleoproteins complex, allow it to incubate for 5 minutes.
3. Add 0.5 mL of chloroform to it, then cap the tube and incubate for 2–3 minutes.
4. 20 minutes of centrifugation at 12,000 × g at 4°C should be sufficient. Separation occurs into a lower phenol-chloroform phase, an interphase, and a colorless upper aqueous phase.
5. Transfer the RNA-containing aqueous phase to a new tube slowly.
6. To the aqueous phase, add 0.5 mL of isopropanol and incubate overnight at −20°C.
7. Centrifuge for 15 minutes at 12,000 × g at 4°C. In the tube, RNA precipitates as a white pellet. Wash the pellet twice using 70% ethanol.
8. Air dry the RNA pellet and resuspend the pellet in 20–50 μl of RNase-free water, by pipetting up and down.

14.4   KEY POINTS

1. If the aqueous phase that contains RNA has color, it is probably due to the presence of a fat layer. Centrifuge the sample before chloroform addition to remove the fat layer.
2. Do not allow the RNA pellet to dry completely.
3. Be sure not to carry any of the organic phase with the RNA sample. It will affect the concentration of RNA. In such a case, precipitate the RNA again with ethanol [3].

4. Be sure not to take any of the interphase (contains the DNA) with the aqueous phase.

14.5   QUIBT 3.0 ANALYSIS

14.5.1   INTRODUCTION

Fluorescent dyes specific to the target are detected by Qubit Fluorometer. Despite their low concentrations, these fluorescent dyes only emit when bound to their targets. Qubit 3.0 Fluorometer and Qubit Flex Fluorometer both measure anything that absorbs at 260 nm, such as DNA, RNA, protein, free nucleotides, or excess salts. A UV spectrophotometry-based measure of DNA and RNA often cannot measure low levels accurately due to its low sensitivity [4]. Target-selective dyes in Qubit assays emit fluorescence when bound to DNA or RNA. As the target molecule is bound to the fluorescent dye in the sample, it emits a signal to show its concentration.

Qubit Fluorometers provide more accurate measurements because the dyes fluoresce only when they come in contact with the target molecule – DNA, RNA, or protein – in your sample. This will prevent you from repeating work because of inaccurate measurements.

14.5.2   MATERIALS

- Qubit® 3.0 Fluorometer
- Thin-wall, clear 0.5 mL PCR tubes
- 200 µl micro pipette
- 200 µl micro pipette tips
- Qubit reagents
- Isolated DNA or RNA

14.5.3   METHODOLOGY

1. Set up assay tubes for the standards and one assay tube for each user sample according to the below-mentioned table.
2. Prepare the Qubit® Working Solution by diluting the Qubit® Reagent 1:200 in Qubit® buffer. Prepare 200 µl of Working Solution for each standard and sample.

	Standard Assay Tubes	User Sample Assay Tubes
Volume of Working Solution (from Step 2) to Add	190 µl	180–199 µl
Volume of Standard (from Kit) to Add	10 µl	–
Volume of User Sample to Add	–	1–20 µl
Total Volume in Each Assay Tube	**200 µl**	**200 µl**

3. Transfer the reaction mix as described in the table to the appropriate thin-wall, clear 0.5 ml PCR tubes, and vortex for 2–3 seconds.
4. At room temperature, incubate the tubes for 2–3 minutes.
5. Insert the tubes in the Qubit® Fluorometer and take the readings.

14.6 KEY POINTS

1. Care should be taken to maintain the tubes at room temperature. Avoid holding the tubes for a longer time, and minimal exposure should be used while taking the Qubit 3.0 reading.
2. The Qubit® assays are very sensitive, and even small amounts of material from a previous sample may result in errors. Use a sterilized PCR tube for readings.
3. Minute bubbles in samples will cause errors in readings. Make sure there are no bubbles in the samples. Centrifugation or slight tapping will often help dispel bubbles.
4. Ensure that the lid is closed while reading standards and samples.
5. Ensure that the standard and sample tubes are filled to 200 μl.

REFERENCES

[1] Chomczynski, P., and Sacchi, N. 1987. Single step method of RNA isolation by acid guanidinium thiocyanate-phenol-chloroform extraction. *Anal. Biochem.* 162, 156–159.
[2] Hummon, A.B., Lim S.R., Difilippantonio, M.J., and Ried, T. 2007. Isolation and solubilization of proteins after TRIzol® extraction of RNA and DNA from patient material following prolonged storage. *BioTechniques* 42, 467–472.
[3] Chomczynski, P. 1993. A reagent for the single-step simultaneous isolation of RNA, DNA and proteins from cell and tissue samples. *BioTechniques* 15, 532–537.
[4] Manchester, K. L. 1996. Use of UV methods for measurement of protein and nucleic acid concentrations. *BioTechniques*, 20, 968–970.

15 Target Gene Amplification and Expression by qRT-PCR

Milan K.L. and Kannan Harithpriya
Department of Biotechnology, School of Bioengineering,
SRM Institute of Science and Technology, Kattankulathur,
Tamil Nadu, India

Dhamodharan Umapathy
Department of Biotechnology, School of Bioengineering,
SRM Institute of Science and Technology, Kattankulathur,
Tamil Nadu, India

Department of Research, Adhiparasakthi Dental College and
Hospital, Melmaruvathur, Tamil Nadu, India

R. Senthilkumar
Department of Biotechnology, School of Applied Sciences,
REVA University, Bangalore, Karnataka, India

K.M. Ramkumar
Department of Biotechnology, School of Bioengineering,
SRM Institute of Science and Technology, Kattankulathur,
Tamil Nadu, India

15.1 INTRODUCTION

A template of RNA is used to synthesize complementary DNA (cDNA) by reverse transcription. A transcription reaction is similar, except that the DNA is the product and mRNA is the template. *In vitro*, components similar to PCR reaction are used for the process. Reverse transcription is performed by combining total mRNA (template) from cells with dNTPs, $MgCl_2$, primers, buffer, and a polymerase enzyme called reverse transcriptase [1]. A cDNA is more stable than RNA that can be quantified with PCR-based analysis. cDNA synthesis is the first step for many protocols in molecular biology because it is the template for downstream applications such as gene expression [2].

Three steps are involved in reverse transcription: primer annealing, DNA polymerization, and enzyme deactivation. Depending on the primer, target RNA, and

DOI: 10.1201/9781003397755-17

55

reverse transcriptase used, the temperature and duration of these steps will differ where polymerization of DNA is the crucial step. Temperature and duration of this step also will depend on the primer choice and reverse transcriptase used. There are differences between reverse transcriptases regarding the thermo-stability, and this subsequently determines the optimal polymerization temperature [1]. It is possible to denature RNA with high GC content or secondary structures by using a thermo-stable reverse transcriptase at a higher reaction temperature (e.g., 50°C) without affecting enzyme activity. An increase in cDNA yield, representation, and length can be attributed to high-temperature incubation procedures. A reverse transcriptase's processivity refers to the number of nucleotides incorporated in a single binding event that determines the amount of time it takes for polymerization reaction [3].

15.2   MATERIALS

- Thermocycler
- Polypropylene micro centrifuge tubes (200 μl)
- Micro pipettes (200, 20, 10 μl)
- Micro pipettes tips (200, 20, 10 μl)
- cDNA Reverse Transcription Kit
  - 10x RT Buffer
  - 10x RT Random Primers
  - 25x dNTP Mix (100 mM)
  - MultiScribe™ Reverse Transcriptase
- Sterile water
- RNAseZap
- Isolated RNA

15.3   METHODOLOGY

1. The input amount of total RNA can be from 100 ng to1 μg of total RNA per 20 μl reaction.
2. Place the components of the kit on ice to thaw.
3. Calculate the required volume of components based on the following table.
4. Prepare the RT master mix on ice as mentioned in the following table.

Component	Volume per Reaction
10x RT Buffer	2.0 μl
25x dNTP Mix	0.8 μl
10x RT Random Primers	2.0 μl
Multiscribe Reverse Transcriptase	1.0 μl
Nuclease-Free Water	4.2 μl
Total Volume	**10 μl**

5. Place the RT master mix on ice and gently mix.
6. Pipette 10 μl of the RT master mix on each tube.

7. Pipette 10 µl of the RNA sample to the RT master mix and pipette up and down to mix.
8. Centrifuge and spin down the contents briefly. Once the tubes are ready for thermal cycling, place them on ice (until the unit is ready to run).
9. Program the thermal cycler as listed in the following table.

	Step 1	Step 2	Step 3	Step 4
Temperature (°C)	25	37	85	4
Time	10 min	120 min	5 min	∞

10. Set the reaction volume to 20 µl.
11. Place the tubes onto the thermal cycler.
12. The cDNA is stored at 4°C until further use. For long-term storage of cDNA, −20°C freezers are used.

15.4   GENE EXPRESSION USING qRT-PCR

15.4.1   INTRODUCTION

Proteins are frequently produced from these genes, but in genes that do not code for proteins, such as rRNA genes and tRNA genes, the product is structural or housekeeping RNA. Additionally, long non-coding RNAs and small non-coding RNAs play key roles in a variety of regulatory processes

Real-time PCR is generally used to study gene expression by examining changes – an increase or decrease – in the expression of a gene-specific transcript. In the investigation, a compound or drug of interest is treated with a defined set of conditions in order to monitor the response of the gene. Several genes can also be studied by looking at their profiles or patterns of expression. Real-time PCR is used by the majority of scientists to perform gene expression analyses, whether they are analyzing quantitative changes in expression levels or looking at overall patterns [4].

Relative quantification allows the comparison of target and reference templates in a sample. The rate of target is measured with respect to a control that is assumed to be invariant. Comparing genetic polymorphism patterns between healthy and diseased tissues is most often measured with relative qPCR [4–6].

SYBR® Green dye is used since it produces fluorescence once it binds to DNA in real-time PCR. In order to amplify the target using this chemistry, two specific primers are needed. Depending on the amount of target generated, dye amounts will vary. Fluorescence emitted by the dye at 520 nm can be detected and related to the amount of target [7,8].

15.4.2   MATERIALS

- QuantStudio 5 real-time PCR instrument
- Quick spin rotor

- Polypropylene micro centrifuge tubes (200 μl)/96 well plate (200 μl)
- Micro pipettes
- Micro pipette tips
- PCR sealing film (optical grade)
- SYBR® Green reagent
- Nuclease-Free Water
- Sample
- Primers
- cDNA or isolated DNA sample

15.4.3 METHODOLOGY

1. Place all components of the reaction on ice.
2. Following the table below, prepare a master mix to run all samples in triplicate.
3. Mix the reagents according to the calculated amount. To account for pipetting errors, add 10% to the volume.
4. Add a triplicate of the "No template Negative Controls" (NTC) along with the reaction.

Component	Volume
SYBR Green Master Mix	10 μl
Forward Primer	1 μl
Reverse Primer	1 μl
cDNA	1 μl
Nuclease-Free Water	7 μl
TOTAL VOLUME	**20 μl**

5. Fill each PCR tube or plate well with 19 μl of the master mix.
6. Add 1 μl of water to the reaction tube for NTC reactions. In the reaction tube, add 1 μl of cDNA solution for experimental purposes.
7. For a few seconds, centrifuge all tubes. Examine each tube or well visually to verify that it contains samples at the bottom with the correct volume.
8. Stir well and spin down the reaction mix.
9. PCR plates or tubes should be sealed and labeled (dependent on the instrument).
10. Labels should not block the instrument excitation or detection light path.

The table below lists how to program a thermal cycler.

	PCR			Melt curve		
	Step 1	Step 2	Step 3	Step 1	Step 2	Step 3
Temperature	95°C	95°C	60°C*	95°C	60°C*	95°C*
Time	5 min	15 sec	35 sec	15 sec	60 sec	15 secs
			35 cycles			
	* Data collection					

15.4.4 KEY POINTS

1. Check for the integrity of RNA prior to cDNA synthesis.
2. Avoid RNase contamination.
3. Include an RNase inhibitor in the reverse transcription setup.
4. In case of low RNA purity, dilute the input RNA in nuclease-free water to reduce the concentration of potential inhibitors.
5. In GC content or secondary structure, denature secondary structures by heating RNA at 65°C for ~5 minutes, then chilling rapidly on ice, prior to reverse transcription.
6. With a gene-specific primer, ensure the primer's sequence is complementary to the 3' end of the target.
7. Appropriate temperature and reaction set up should be followed for the selected RT.

REFERENCES

[1] Klickstein, L. B., Neve, R. L., Golemis, E. A., and Gyuris, J. Conversion of mRNA into double-stranded cDNA. *Current Protocols Mol. Biol.* 29 (1995).
[2] Bustin, S.A., Quantification of mRNA using real-time reverse transcription PCR (RT-PCR): Trends and problems, *J. Mol. Endocrinol.* 29, 23–29 (2002).
[3] Rees, W., et al., Betaine can eliminate the base pair composition dependence of DNA melting, *Biochemistry*, 32, 137–144 (1993).
[4] Morrison, T.B., et al., Quantification of low-copy transcripts by continuous SYBR® Green I monitoring during amplification, *BioTechniques*, 24, 954–962 (1998).
[5] Sambrook, J., et al. *Molecular cloning: A laboratory manual*, 3rd Edition, Cold Spring Harbor Laboratory Press, New York (2000). Catalog Number M8265.
[6] Lovatt, A., et al., Validation of Quantitative PCR Assays, *BioPharm.*, March 2002, pp. 22–32.
[7] Dieffenbach, C., and Dveksler, G., (eds.) *PCR Primer: A Laboratory Manual*, Cold Spring Harbor Laboratory Press, Cold Spring Harbor, NY (1995). Catalog Number Z364118.
[8] Kellogg, D.E., et al., TaqStart Antibody: "hot start" PCR facilitated by a neutralizing monoclonal antibody directed against Taq DNA polymerase, *Bio Techniques*, 16, 1134–1137 (1994).

16 Cell Lysis

Protein Extraction Using RIPA Buffer

Kannan Harithpriya and Ravichandran Jayasuriya
Department of Biotechnology, School of Bioengineering,
SRM Institute of Science and Technology, Kattankulathur,
Tamil Nadu, India

Dhamodharan Umapathy
Department of Biotechnology, School of Bioengineering,
SRM Institute of Science and Technology, Kattankulathur,
Tamil Nadu, India

Department of Research, Adhiparasakthi Dental College and
Hospital, Melmaruvathur, Tamil Nadu, India

K.M. Ramkumar
Department of Biotechnology, School of Bioengineering,
SRM Institute of Science and Technology, Kattankulathur,
Tamil Nadu, India

16.1 INTRODUCTION

Protein extraction is the foremost step in many biochemical and analytical techniques. Tissue or cell homogenization plays a crucial role in determining the yield of the protein. Cell lysis can be done in various methods, but the most widely used are chemical lysis and enzymatic lysis combined with sonication. Among the various chemical lysis methods, RIPA buffer tends to be efficient in lysing the cell and preventing protein degradation. This buffer is widely used for efficient total cell lysis and is highly effective in the solubilization of proteins from cultured mammalian cells. RIPA lysis buffer comprises both ionic and non-ionic detergents capable of extracting proteins from membrane structures and from cytoplasmic and nuclear membranes. The salts in the lysis buffer (Tris-HCL and NaCl) regulate and maintain the pH and osmolarity. Triton X-100, a non-ionic surfactant, disrupts the polar head group present in the lipid bilayer of the cells, which in turn destroys the integrity and increases the permeability of the membrane. The addition of ionic detergents sodium deoxycholate and SDS to the lysis buffer helps extract and

DOI: 10.1201/9781003397755-18

solubilize the protein of interest. Finally, phenyl methyl sulfonyl fluoride (PMSF), a non-specific protease inhibitor, reacts with serine residues to inhibit trypsin, chymotrypsin, and papain against digestive functions of protease [1].

16.2 MATERIALS

- 1X RIPA lysis buffer (stored at 4°C)
  - 50 mM Tris HCl (pH 7.4)
  - 150 mM NaCl
  - 0.5% Sodium deoxycholate
  - 0.1% SDS
  - 1 mM EDTA
  - 1% Triton X-100
- (Add 1 mM phenyl methyl sulfonyl fluoride to the ice-cooled lysis buffer.)
- Microfuge tubes
- Micropipettes and tips
- Distilled water
- Adherent or monolayer cell culture
- Phosphate buffer saline
- Cell scraper

16.3 METHODOLOGY

Protein extraction from cultured cells:

1. **For monolayer adherent cells:** Once 70% of the confluency is attained, rinse the monolayer cells twice with ice-cooled PBS. With the help of a cell scraper, scrape out and transfer the cell suspension to the fresh tube, and centrifuge at 1500 RPM for 10 minutes at 4°C. Aspirate the supernatant completely without disturbing the pellet.
   (After trypsinization, proceed to step 3 for cell lysis)
2. **For suspension cells:** The cell suspension is centrifuged at 1500 RPM for 10 minutes, the supernatant is removed completely. Wash the pellet with ice-cooled PBS twice.
3. Add 100 μL of ice-cooled cell lysis buffer to the pellet and incubate on ice for 15 minutes.
4. Vortex the sample thoroughly and sonicate with 30 sec/30 secs on and off for up to 10 cycles at 4°C until the cell is completely homogenized.
5. Transfer the supernatant to the new tube and centrifuge at 12,000 RPM for 10 minutes at 4°C and store the supernatant at −20°C until further use [2].

16.4 TROUBLESHOOTING

1. If total protein yield is low, ensure the cell pellet is thoroughly suspended in RIPA buffer and increased incubation time.

2. Use a minimal amount of lysis buffer to avoid low concentration of proteins.
3. It is important to add protease inhibitors to the buffer before use to eliminate proteolysis [3].

REFERENCES

[1] https://www.thermofisher.com/in/en/home/life-science/protein-biology/protein-biology-learning-center/protein-biology-resource-library/pierce-protein-methods/overview-cell-lysis-and-protein-extraction.html accessed on 25th July 2021.
[2] Protein extraction from Tissues and Cultured Cells using Bioruptor® Standard & Plus- Diagenode – Innovative Epigenetic Solution.
[3] https://www.embl.de/pepcore/pepcore_services/protein_purification/extraction_clarification/cell_lysates_ecoli/. Accessed on 25 July 2021.

17 Estimation of Protein by Bradford's Assay

Kannan Harithpriya and Ravichandran Jayasuriya
Department of Biotechnology, School of Bioengineering,
SRM Institute of Science and Technology, Kattankulathur,
Tamil Nadu, India

Dhamodharan Umapathy
Department of Biotechnology, School of Bioengineering,
SRM Institute of Science and Technology, Kattankulathur,
Tamil Nadu, India

Department of Research, Adhiparasakthi Dental College and
Hospital, Melmaruvathur, Tamil Nadu, India

K.M. Ramkumar
Department of Biotechnology, School of Bioengineering,
SRM Institute of Science and Technology, Kattankulathur,
Tamil Nadu, India

17.1 INTRODUCTION

During protein purification, protein concentration plays an important role in determining enzyme activity and preparing protein samples for electrophoresis and western transfer. Among various methods, the Bradford assay is used to quantify the total protein present in the sample. Most of the protein assay determines the reaction between the dye and sample of interest that will in turn increase the absorbance. Basically the more protein is present, the higher will be the absorbance.

The Bradford protein estimation assay is the quick and simple method to determine protein concentration. The dye Coomassie Brilliant Blue G-250(CBB) has two sulfonic acid groups and six phenyl groups that interact with protein through positive charge and hydrophobic residues. Under the strong acidic condition, the CBB dye is stable as a doubly protonated red form. In contrast, upon binding to protein, it is most stable as unprotonated and gives rise to a blue color complex. This interaction forms a protein-dye complex that increases the absorbance of CBB from 465 to 595 nm. The most commonly used protein standard is Bovine serum albumin (BSA). The absorbance of the known protein standard is obtained, and a linear regression graph is plotted against the concentration of the known sample. An unknown protein concentration can then be estimated by plotting against the standard curve.

DOI: 10.1201/9781003397755-19

17.2 MATERIALS

- BSA stock solution (2 mg/ml)
- Bradford reagent – Coomassie Brilliant Blue G-250
- Test tubes or 96 well flat bottom plates
- Distilled water

17.3 METHODOLOGY

1. Take seven different concentrations of 2 mg/ml of BSA [0.2–2 mg/ml] in individual test tubes and make up the total volume of 100 µl by adding distilled water, as mentioned in the table that follows.
2. Add the 100 µl unknown protein sample to the fresh tube.
3. To each tube add 2 ml of Bradford's reagent and incubate at room temperature for 5–45 minutes.
4. The absorbance is measured at 595 nm in a spectrophotometer; the tube with water and Bradford serves as a blank.
5. The graph is then plotted against the concentration of known protein on the x-axis and absorbance at 595 nm on the y-axis. Obtain the linear regression equation y = mx + c to determine the unknown protein concentration.

(**Note:** Always run the sample in duplicates or triplicates and use the mean absorbance to get the standard curve graph and equation. The linear regression equation can be obtained by plotting the graph in MS Excel, discussed next.)

17.4 PLOTTING THE GRAPH IN MS EXCEL

- Preload the data in an Excel sheet. Select the data that are entered, and insert a scatter plot.
- Right-click at any point on the blue line displayed on the graph, and select "add trend lines."
- Select the linear option on trend/regression type, and at the end check the box to both display equation on chart and display R-squared value on chart. The linear regression value will then be displayed on the graph from which the concentration of unknown can be determined by substituting the absorbance value in y [1,2].

S.No	Proteins		Volume of Distilled water (µl)	Volume of bradford reagent (ml)		Absorbance at 595 nm
	(µg)	(µl)				
1	Blank	–	100	2		
2	0.2	10	90	2		
3	0.4	20	80	2	Incubate at room temperature for 5–45 minutes	
4	0.6	30	70	2		
5	0.8	40	60	2		
6	1.0	50	50	2		
7	1.5	75	25	2		
8	2.0	100	0	2		
9	Unknown	100 µl	–	2		

REFERENCES

[1] Bradford, M.M. (1976) A rapid and sensitive method for the quantitation of micro-gram quantities of protein utilizing the principle of protein-dye binding. *Anal. Biochem.* **72**, 248–254.
[2] Sedmak, J.J. and Grossberg, S.E. (1977) A rapid, sensitive and versatile assay for protein using Coomassie Brilliant Blue G250. *Anal. Biochem.* **79**, 544–552.

18 SDS-PAGE Analysis

Kannan Harithpriya and Goutham V. Ganesh
Department of Biotechnology, School of Bioengineering,
SRM Institute of Science and Technology, Kattankulathur,
Tamil Nadu, India

Dhamodharan Umapathy
Department of Biotechnology, School of Bioengineering,
SRM Institute of Science and Technology, Kattankulathur,
Tamil Nadu, India

Department of Research, Adhiparasakthi Dental College and
Hospital, Melmaruvathur, Tamil Nadu, India

K.M. Ramkumar
Department of Biotechnology, School of Bioengineering,
SRM Institute of Science and Technology, Kattankulathur,
Tamil Nadu, India

18.1 INTRODUCTION

Electrophoresis is said to be the process of migration of charged molecules with respect to the applied electric field. SDS-PAGE (sodium dodecyl sulfate-polyacrylamide gel electrophoresis) is an analytical technique used to separate proteins from complex mixtures based on their relative molecular mass and charge. The migration rate of samples depends on the size, shape of the molecule, and their net charge with respect to the electric field gradient. An anionic detergent SDS strongly binds to the protein and denatures the protein to produce linear polypeptide chains. On average, one SDS molecule binds to two amino acids. The presence of β-mercaptoethanol helps in protein denaturation by reducing all disulfide bonds. The SDS-protein complex carries a net negative charge, which uniformly migrates toward the anode and separates based on their molecular mass. In SDS-PAGE, two different gels are cast, a separating gel topped by stacking gel varied by pH of the buffer. The co-polymerization of acrylamide monomers with cross-linking reagent N-N'methylene bisacrylamide forms a clear transparent gel. This is said to be the example of free radical catalysis and is initiated by APS and catalyst TEMED. Hereupon TEMED addition to the solution the decomposition of per sulfate ion occurs, giving rise to the free radical. This free radical reacts with the Bis-acrylamide and forms a mesh-like structure with the pores through which the protein sample can migrate [1].

DOI: 10.1201/9781003397755-20

The glycine in the stacking gel remains in neutral zwitterionic form, wherein chlorine ion tends to be an effective current carrier. The SDS-coated protein molecules and dye have a charge-to-mass ratio greater than glycine. The complex concentrates into a thin band between the chlorine ions and the glycine molecules at the interface between stacking and separating gels. The negatively charged protein-SDS move according to their relative mobilities and separate gel according to relative molecular mass. The resolved protein in the gel can be stained using silver stain or using Coomassie Brilliant Blue. Comparatively, silver stain tends to be more effective and sensitive than Coomassie stain [2].

18.2   MATERIALS

- 30% Acrylamide/bisacrylamide solution
- 1.5 M Tris-HCL pH 8.8 (separating gel)
- 0.5 M Tris-HCL pH 6.8 (stacking gel)
- 10% Ammonium per sulfate
- 10% SDS
- N,N,N,N'-tetramethylethylenediamine (TEMED)
- 10X running buffer pH 8.3
  - 25 mM Tris-HCl
  - 200 mM Glycine
- 0.1% SDS
- Staining solution
  - 0.1% CBB G-250
  - 40% methanol
  - 10% Glacial acetic acid
    - (Filter the solution before use)
- Destaining solution
  - 50% methanol
  - 35% Glacial acetic acid
    - (Make up to 100 ml using distilled water)
- 6x Laemmli loading buffer
  - 0.004% bromophenol blue
  - 20% glycerol
  - 4% SDS
  - 0.125 M Tris-HCl
  - 10% 2-mercaptoethanol
- SDS-PAGE gel apparatus

18.3   METHODOLOGY

1. Set the gel casting plate on the stand, ensure it is sealed completely, and check for any leakage using distilled water. Blot dry the plate.
2. Prepare the separating gel without adding the polymerizing agent.

Components	8%	10%	12%	15%
1.5 M Tris-HCL pH 8.8	1.3 ml	1.3 ml	1.3 ml	1.3 ml
Distilled water	2.3 ml	1.9 ml	1.6 ml	1.1 ml
30% acryl/Bis-acryl	1.3 ml	1.7 ml	2 ml	2.5 ml
10% SDS	50 μl	50 μl	50 μl	50 μl
10% APS	50 μl	50 μl	50 μl	50 μl
TEMED	3 μl	2 μl	2 μl	2 μl

3. Cast the gel solution in the three-fourth of the plate assembled, and overlay the surface with methanol or isopropanol to avoid uneven casting.
4. Allow the gel to set completely for 30 minutes at room temperature and 2 ml of 5% stacking gel is prepared.

- 0.5 M Tris-HCl (pH 6.8)   : 0.25 ml
- 30% acryl/Bis   : 0.33 ml
- Distilled water   : 1.4 ml
- 10% SDS   : 20 μl
- 10% APS   : 20 μl
- TEMED   : 2 μl

(Note: Always add TEMED just before pouring the gel onto the plate.)
5. After 30 minutes, discard the overlayed water and add 5% stacking gel solution until it overflows.
6. Insert the comb immediately and ensure no air bubbles are trapped in the gel or near the wells. Set it aside for 30 minutes.
7. To the volume of the protein sample, add an equal volume of loading buffer and denature it at 95°C for 5 minutes.
8. After polymerization is complete, carefully remove the comb and assemble the gel cassette in the electrophoresis tank. Add 1X running buffer to the top and bottom reservoir.
9. Load the appropriate volume of denatured protein sample and the protein ladder of interest into the well.
10. Connect the power pack, and allow the gel to run at 80 V until the protein stack starts separating gel and, run at 100 V once it reaches the separating gel. Allow the gel to run until the dye front reaches the bottom of the resolving gel.
11. After completing electrophoresis, remove the plate carefully, and stain the gel using gel staining solution for a minimum of 2 hours.
12. Destain the gel by soaking in destaining solution on a rocking platform for 4–7 hours or overnight, and the bands can be visualized using a gel documentation system.

18.4   PRECAUTIONS

1. Acrylamide monomer is said to be a neurotoxin; care should be taken while handling the chemical. Use gloves, and avoid inhaling it.
2. Coomassie Brilliant Blue is harmful; wear gloves while using it.

18.5   TROUBLESHOOTING

1. Always prepare fresh APS solutions to enhance polymerization.
2. Since beta mercaptoethanol is very much essential in breaking the disulfide bond, it is advised to prepare fresh during the day of the experiment.
3. Refrigerate acrylamide/bi-acrylamide solution, APS, and TEMED.
4. If weak/no bands are seen after staining the gel, it may be due to a low concentration of protein or proteins that have been degraded. To avoid this, increase the concentration of the sample accordingly and the addition of protease inhibitor will eliminate the degradation.
5. Run the gel with minimum voltage, to avoid poor band resolution.
6. If the protein sample interest size is unknown, it is preferred to use 5–15% gradient gel.
7. Add the appropriate concentration of protein to eliminate protein smearing.
8. If the proteins are hydrophobic, they may precipitate in the well. To avoid precipitation, about 4 M of urea can be added to the sample.
9. Minimum time should be taken between the sample application and power run to exclude the spreading of lateral bands.
10. The smile effect in the gel is caused due to the high heat at the center of the gel at the ends. If such an effect is detected, decrease the voltage setting and use fresh buffers [3].

REFERENCES

[1] Abe, T., Naito, T., Uemura, D. Sodium dodecyl sulfate-polyacrylamide gel electro-phoresis (SDS-PAGE) analysis of palytoxin. *Natural Product Communications.* August 2017. 10.1177/1934578X1701200815
[2] Manns, J.M. (2011). SDS-polyacrylamide gel electrophoresis (SDS-PAGE) of proteins. *Current Methodologies in Microbiology*, 22: A.3M.1–A.3M.13. 10.1002/97804 71729259
[3] https://ruo.mbl.co.jp/bio/e/support/method/sds-page.html (accessed on 1 August 2021).

19 Immunoblotting

Kannan Harithpriya and Karan Naresh Amin
Department of Biotechnology, School of Bioengineering,
SRM Institute of Science and Technology, Kattankulathur,
Tamil Nadu, India

Dhamodharan Umapathy
Department of Biotechnology, School of Bioengineering,
SRM Institute of Science and Technology, Kattankulathur,
Tamil Nadu, India

Department of Research, Adhiparasakthi Dental College and
Hospital, Melmaruvathur, Tamil Nadu, India

K.M. Ramkumar
Department of Biotechnology, School of Bioengineering,
SRM Institute of Science and Technology, Kattankulathur,
Tamil Nadu, India

19.1 INTRODUCTION

Western blotting, widely known as western transfer, is a rapid analytical technique for identifying the specific protein of interest in a complex mixture of protein samples separated based on relative molecular mass by SDS-PAGE gel electrophoresis. This technique is said to be an immunoblotting technique that uses specific antibodies that attach to the target of interest that will migrate to the membrane. The membrane widely used to undergo western transfer is PVDF (polyvinylidene fluoride) or nitrocellulose membrane. The separated protein of interest is then blotted onto the membrane, and the specific targeted protein is then stained with the conjugated primary and secondary antibodies labeled with enzymes like horseradish peroxidase. By visualizing and analyzing the intensity and location on the membrane, the protein expression can be determined. Western blot technique can detect protein as low as 1 ng with high-resolution capacity, specificity, and sensitivity.

19.2 MATERIALS

- Western transfer cassette and accessories
- Nitrocellulose membrane or PVDF membrane
- Western transfer buffer
  - 12 mM Tris Base

DOI: 10.1201/9781003397755-21

- 95 mM glycine
- 20% methanol
     (Make up the volume to 1 liter with distilled water.)
- Blocking solution (3% BSA)
- TBST (10X stock solution)
  - 200 mM Tris
  - 1.5 M NaCl
       (Make up the final volume to 1 liter with distilled water.)
- Primary antibody
- Enzyme-conjugated secondary antibody
- SDS-PAGE apparatus

19.3 METHODOLOGY

1. The western transfer cassette is set up so that it is completely submerged in the transfer buffer.
   - To prepare the western transfer cassette, submerge the cassette in the transfer buffer tray and place the sponge pad on the cassette for support and care to be taken to avoid air bubbles. The small piece of blotting paper is placed above the sponge cassette
   - Carefully remove the gel from the casting plate, place the pre-soaked cassette and pre-soaked nitrocellulose membrane, and carefully place the gel on top of the membrane.

   (*Note: Proper care should be taken to avoid air bubbles.*)

   - On top of the gel, place the blotting papers or sponge pads so that a sandwich is formed firmly.
2. Place the cassette into the tank filled with western transfer buffer, connect the assembly to a power supply, and allow the transfer to happen for 1.5 hours at the constant volt of 400 mA.
3. For the SEMI-DRY transfer method, place the sandwich on the anode surface, roll the cassette assembly to avoid air bubbles, place the cathode on top, and close the semi-dry apparatus. The transfer occurs for 30–45 minutes at 25 V.

19.4 DEVELOPMENT/IMMUNOBLOTTING

1. Once the transfer is over, the membrane is then blocked with 3% blocking solution for 30 minutes with gentle shaking.
2. After 30 minutes, wash the membrane twice with 1X TBST buffer. (*Note: For 1 X: 1 part of 10x stock with 9 parts of water.*)
3. The diluted primary antibody with desired concentration is then added and incubated for 4 hours at room temperature. (*Note: The concentration of primary antibody depends on the experiment; the optimum incubation time for the primary antibody must be predetermined.*)
4. After incubation, wash the membrane twice with 1X TBST buffer.

5. The membrane is then incubated with enzyme-conjugated secondary antibody for 1–2 hours at room temperature.
6. After incubation, wash the membrane twice with 1X TBST buffer and analyze it for chemiluminescence with an ECL substrate kit.
7. The pattern for protein is visualized through the ChemiDoc **MP** System, where the band position of the protein of interest is compared to that of the protein ladder to determine the protein size.

19.5 TROUBLESHOOTING

1. Care should be taken during the washing step, to eliminate the faint band or no signal during detection.
2. Minimum exposure time is preferred while imaging the blot to avoid no signal.
3. Always run a positive control to detect your protein of interest.
4. If multiple bands are seen at low molecular weight, the protein may have been digested by the enzyme proteases. To prevent protein degradation, addition of a protease inhibitor is necessary.
5. If multiple bands are detected at high molecular weight, protein may have formed multimers due to the presence of a disulfide bond. To prevent the formation of multimers, it is advisable to boil the sample for a longer period of time in Laemmli buffer solution during sample preparation.
6. While performing band detection, if too many bands are visible at various MW, it indicates that the cell line used for the study may have passaged multiple times, which may elicit the difference in their protein expression profile. Instead, run an original non-passaged cell line for cross-verification.
7. Acrylamide percentage in the gel plays a major role. If the percentage is high, we may encounter very high bands. Always for high protein concentration, the acrylamide percentage should be low.
8. It is suggested to alter the migration voltage and temperature during the transfer as per the requirement to avoid uneven protein bands upon detection.
9. Care should be taken while polymerizing the gel. If not polymerized completely, the bands of protein may appear uneven and smudged [1,2].

REFERENCES

[1] Mahmood, T., Yang, P.C. Western blot: technique, theory, and trouble shooting. *N Am J Med Sci.* 2012;4(9):429–434. 10.4103/1947-2714.100998
[2] https://www.antibodies-online.com/resources/17/1224/western-blotting-immunoblot-gel-electrophoresis-for-proteins/ (accessed on 4 August 2021).

Section 3

Cell Staining Techniques

20 Cell Nuclear Staining by DAPI

Ravichandran Jayasuriya, Karan Naresh Amin, and Sundhar Mohandas

Department of Biotechnology, School of Bioengineering, SRM Institute of Science and Technology, Kattankulathur, Tamil Nadu, India

R. Senthilkumar

Department of Biotechnology, School of Applied Sciences, REVA University, Bangalore, Karnataka, India

K.M. Ramkumar

Department of Biotechnology, School of Bioengineering, SRM Institute of Science and Technology, Kattankulathur, Tamil Nadu, India

20.1 INTRODUCTION

Nuclear staining is one of the basic experiments to visualize the live nucleus inside the cell. All cells will have a nucleus that is transparently visible in a phase contrast microscope. However, to confirm the localization and shape of the nucleus, nuclear staining is used. The nuclear staining dye directly binds to the nucleic acid of a live cell. DAPI and Hoechst are two majorly used dyes, which are termed as minor-groove binding dyes, that are non-toxic to cells. These dyes can be used to stain the nucleus in a viable cell as well as in a fixed cell. DAPI binds to the AT-rich regions of DNA as it has a higher affinity than that of GC-rich DNA. Moreover, DAPI is more stable at 4°C and has a slightly higher photostability than Hoechst.

20.2 MATERIALS

- Cell culture facility
- Six-well cell culture plates
- Pipettes and micro tips (sterile)
- Tabletop centrifuge
- DAPI dye
- Phosphate buffer saline (PBS)
- 3.7% formaldehyde (for cell fixation)

DOI: 10.1201/9781003397755-23

- Mounting medium
- 0.2% Triton X-100

20.3   REAGENT/DYE PREPARATION

1. The stock solution may be prepared for 10 mg/ml in sterile water (*Note: DAPI is more soluble in water than in PBS.*) Prepare a working concentration (1:5000) right before use for best results.
2. 3.7% formaldehyde can be prepared by adding 3.7 ml of formaldehyde to 96.3 ml of water.

20.4   METHODOLOGY

1. The cells are grown on a coverslip placed inside a six-well culture plate.
2. Once the cells reach confluence, aspirate the spent medium and wash the cell with 1X PBS (3 times).
3. Cells are fixed with 3.7% formaldehyde and incubated for 10 minutes at RT.
4. Formaldehyde is aspirated, and the cells are washed with 1X PBS (3 times) to remove residual formaldehyde.
5. To permeablise the cells, 1–2 ml of 0.2% Triton X-100 is added and incubated for 5–10 minutes.
6. Triton X-100 is then aspirated, and the cells are washed with 1X PBS (3 times).
7. The cells are then incubated with DAPI (1:5000) in PBS for 10–15 minutes.
8. DAPI is aspirated, and the cells are washed with 1X PBS (3 times).
9. Mount the coverslips with mounting medium (optional).
10. The cells can be imaged at an excitation wavelength of ~359 nm and emission wavelength of ~461 nm.

20.5   INTERPRETATION OF THE IMAGE

The live nucleus looks blue in color when imaged at the preferred wavelength [1,2].

REFERENCES

[1] T. Pommerencke, T. Steinberg, H. Dickhaus, P. Tomakidi, and N. Grabe. Nuclear staining and relative distance for quantifying epidermal differentiation in biomarker expression profiling. *BMC Bioinformatics*. 9:473 (2008).
[2] B.I. Tarnowski, F.G. Spinale, and J.H. Nicholson. DAPI as a useful stain for nuclear quantitation. *Biotechnic & Histochemistry: Official Publication of the Biological Stain Commission*. 66:297–302 (1991).

21 Immunofluorescence Staining

Ravichandran Jayasuriya, Kannan Harithpriya, and Sundhar Mohandas
Department of Biotechnology, School of Bioengineering, SRM Institute of Science and Technology, Kattankulathur, Tamil Nadu, India

R. Senthilkumar
Department of Biotechnology, School of Applied Sciences, REVA University, Bangalore, Karnataka, India

K.M. Ramkumar
Department of Biotechnology, School of Bioengineering, SRM Institute of Science and Technology, Kattankulathur, Tamil Nadu, India

21.1 INTRODUCTION

To analyze proteins in different tissue sections, cultured cell lines, or individual cells, the technique of immunofluorescence is used. This technique makes use of various fluorophores in order to visualize the target protein. It is based on antigen–antibody binding, where the fluorophore is tagged with the antibody that recognizes the antigen (i.e., target protein). This allows the visualization of the target protein inside the cell under a fluorescence or a confocal microscope. This technique can also be used to visualize cytoskeletons or intermediate filaments in cells or in tissue sections. Immunofluorescence can be combined with DAPI stain to distinguish and label the nucleus.

The antibodies used for this technique are conjugated to fluorophores of wide intensity. In the case of direct detection, the primary antibody is covalently coupled to a fluorophore. This recognizes the target antigen and emits fluorescence when excited at its specific wavelength. This reduces the risk of cross-reactivity and minimizes the steps followed with indirect staining, i.e., through a secondary antibody. In indirect immunofluorescence, the secondary antibody that carries the fluorophore binds with the unconjugated primary that is specific to the antigen in the samples. However, this methodology is more complex; it allows the use of various secondary and fluorophores of choice. Both these methods indulge in blocking of non-specific targets using BSA and washing using Triton X-100, as seen in western blotting.

DOI: 10.1201/9781003397755-24

21.2 MATERIALS

- Cell culture facility
- Six-well cell culture plates
- Glass slides and coverslips
- Pipettes and micro tips (sterile)
- Phosphate buffer saline (PBS)
- 4% paraformaldehyde
- 0.1% Triton X-100
- 1% BSA
- Primary antibody (either direct conjugated or primary- and secondary-conjugated antibody)
- Mounting medium

21.3 REAGENT PREPARATION

1. 4% paraformaldehyde can be prepared by adding 4 mL of 100% formaldehyde in 96 mL of millipore water (scale to desired volume accordingly).
2. 1% BSA can be prepared by adding 1 g of BSA in 100 mL of millipore water (scale to desired volume accordingly).
3. 0.1% Triton X-100 can be prepared by adding 0.1 mL of 100% Triton X-100 to 100 mL of 1X PBS (scale to desired volume accordingly).
4. For antibody dilution, please refer to the manufacturer's instructions. Usually, 1:50 is preferred for direct conjugated antibodies, and differs on using a secondary-conjugated antibody.

21.4 METHODOLOGY

1. In the case of cells, plate approximately 0.2 million cells on the coverslips and incubate in the appropriate medium with 10% FBS overnight (*Note: The coverslips can be sterilized by placing them in UV light on a BSC for about 20 minutes. Depending on the cell type, Poly-L-lysine coating may be given prior to seeding.*)
2. Once the cells are confluent, aspirate the medium carefully and add the appropriate medium with 0.5–1% FBS with the treatment compound. (*Note: Cells may detach if not properly bound or when the coverslip slides.*)
3. In the case of tissues, they are directly washed with 1X PBS (2 times).
4. Fix the samples by adding 4% formaldehyde, and incubate for about 15 minutes at room temperature.
5. Aspirate formaldehyde followed by washing with 1X PBS (2 times).
6. Permeabilize the samples by adding a sufficient amount of 0.1% Triton X-100 and incubate for 10 minutes. (*Note: This step is not required in the case of staining a surface epitope.*)
7. Aspirate triton X-100 and wash with 1X PBS (2 times).

8. The samples are then blocked for 1–2 hours using 1% BSA, followed by washing with 1X PBS (2–3 times).

9. Incubate the samples with fluorescence-conjugated primary antibody at 4°C overnight in the dark. (*Note: In the case of indirect staining, use overnight incubation of primary followed by wash with 1X PBS and then secondary conjugated antibody at room temperature for 2–3 hours.*)

10. After antibody incubation, the samples are again washed with 1X PBS (3–5 times).

11. To counterstain the nucleus, DAPI may be used, as mentioned in the previous section, followed by washing (Optional).

12. The samples are then mounted using a drop of anti-fade mounting medium.

13. Store the slides in the dark (preferably at 4°C) until visualizing them on a fluorescence or in a confocal microscope.

21.5   INTERPRETATION OF THE EXPECTED RESULTS

The images can be observed using fluorescence or a confocal microscope. The nucleus can be located upon DAPI staining in blue color. Based on the excitation of fluorophore, the color intensity of the target varies [1,2].

REFERENCES

[1] K. Im, S. Mareninov, M.F.P. Diaz, and W.H. Yong. An Introduction to Performing Immunofluorescence Staining. *Methods in Molecular Biology.* 1897: 299–311 (2019).

[2] S. Zaqout, L.L. Becker, and A.M. Kaindl. Immunofluorescence Staining of Paraffin Sections Step by Step. *Frontiers in Neuroanatomy.* 14: 582218 (2020).

22 Estimation of Intracellular ROS using H₂DCFDA Molecular Probes

Ravichandran Jayasuriya, Karan Naresh Amin, Sundhar Mohandas, and K.M. Ramkumar
Department of Biotechnology, School of Bioengineering, SRM Institute of Science and Technology, Kattankulathur, Tamil Nadu, India

22.1 INTRODUCTION

Mitochondrion is an important membrane-bound cell organelle that generates energy that is much needed for the cellular biochemical precesses to take place. The generated energy is stored as small molecules called adenosine triphosphate (ATP) and is released when needed – hence, the name "power house of the cell." Also, mitochondrial oxidative phosphorylation releases reactive oxygen species (ROS) such as superoxide anion (O_2.), hydrogen peroxide (H_2O_2), and hydroxyl radical (HO•) consisting of radical and non-radical oxygen species, which are formed by the partial reduction of oxygen. Apart from mitochondrial phosphorylation, interactions with xenobiotic compounds also release ROS. Hence, it is evidenced that all cells generate ROS at most of the times during their aerobic metabolisms.

ROS plays a protective and functional role, for example, the maturation of dendritic cells. But more often the overproduction of ROS or decreased cellular antioxidant defense to counteract the generated ROS ends up in the onset of oxidative stress. This oxidative stress is implicated with various pathological diseases by damaging nucleic acids, proteins, and lipids. For example, tumor growth and metastasis have been found to be strongly associated with the amount of ROS produced.

The structure and function of proteins are disturbed by ROS as follows. The redox-reactive cysteine residues (Cys) in the protein undergo oxidation and form sulfenic acid (–SOH). This sulfenic acid is ready to form disulfide bonds (–S–S–) and undergoes further oxidation to form sulfinic (–SO_2H) or sulfonic (–SO_3H) acid. Also, it tends to form sulfonamide when nitrogen is available.

2′,7′-Dichlorodihydrofluorescein diacetate (H₂DCFDA) is a widely used molecular probe that is used to estimate the cellular ROS. H₂DCFDA readily

DOI: 10.1201/9781003397755-25

diffuses into the cell and is converted to 2′,7′-dichlorodihydrofluorescein (H$_2$DCF) in the presence of esterases. H$_2$DCF is then oxidized to 2′,7′-dichlorofluorescein (DCF) in the presence of ROS. This reaction emits a strong fluorescence that is measurable at 498 and 522 nm of excitation and emission, respectively. This measurement includes free radicals such as H$_2$O$_2$, HO•, ONOO–.

22.2 MATERIALS

- Cell culture facility
- six well cell culture plates
- Pipettes and micro tips (sterile)
- Tabletop centrifuge
- 2′,7′-Dichlorodihydrofluorescein diacetate (H$_2$DCFDA)
- Dimethylsulfoxide (DMSO)
- Phosphate buffer saline (PBS)
- Hydrogen peroxide (H$_2$O$_2$)

22.3 DYE PREPARATION

H$_2$DCFDA is to be prepared right before use for better results. The stock solution may be prepared for 10 mM, for which 50 μg of dye is dissolved in 8.6 μl of DMSO. Store the dye in the dark at the appropriate temperature.

22.4 METHODOLOGY

1. Plate approximately $0.2 \times 10 \times 10^6$ cells per well in the appropriate medium with 10% FBS/FCS on a sterile six-well culture plate.
2. Once the cells reach confluence, change the medium to no serum or low serum (0.5–1% FBS) for at least 2 hours prior to treatment.
3. The medium is aspirated, and fresh medium with 10–50 μM H$_2$O$_2$ is added, followed by an overnight incubation in a CO$_2$ incubator. (*Note: Addition of dye with more serum may cause de-esterification of DCF.*)
4. The spent medium is removed, and the cells are washed with 1X PBS (2 times).
5. Aspirate PBS, and add fresh medium with reduced serum.
6. To the fresh medium, directly add 1–10 μM H$_2$DCFDA from the stock prepared. (*Note: Over concentration of H$_2$DCFDA will result in artifactual photochemical oxidation.*)
7. Incubate the cells with dye for 20–30 minutes in the dark on an incubator.
8. Remove the medium, and wash the cells with 1X PBS.
9. Add 300 μl of trypsin-EDTA to each well, and incubate for at least 3–5 minutes.
10. Confirm the cell detachment visually using an inverted microscope.

11. Add 1 ml of fresh medium with 10% FBS to inactivate trypsin.
12. Collect the cells by centrifugation.
13. Wash the pelletted cells using 1X PBS, suspend them in 0.5–1 ml of 1X PBS and analyze using a flow cytometer.

22.5   INTERPRETATION OF EXPECTED RESULTS

The results will be in the form of peaks, and analysis of the results is to be based on the peak shift. The extent of the peak shift towards M2 (right side) indicates the cells with the extent of ROS. The control cells (untreated) will have their peak at M1 (left side). As the concentration of H_2O_2 increases, the shift in peak towards M2 can be observed. In this way, the cells with 50 μM H_2O_2 will have a peak shift towards M2, indicating higher concentration of intracellular ROS [1–3].

REFERENCES

[1] R. Ameziane-El-Hassani, M. Boufraqech, O. Lagente-Chevallier, U. Weyemi, M. Talbot, D. Metivier, F. Courtin, J.M. Bidart, M. El Mzibri, M. Schlumberger, and C. Dupuy. Role of H_2O_2 in RET/PTC1 chromosomal rearrangement produced by ionizing radiation in human thyroid cells. *Cancer Research*. 70:4123–4132 (2010).

[2] E. Eruslanovand and S. Kusmartsev. Identification of ROS using oxidized DCFDA and flow-cytometry. *Methods in Molecular Biology*. 594:57–72 (2010).

[3] M. Roesslein, C. Hirsch, J.P. Kaiser, H.F. Krug, and P. Wick. Comparability of in vitro tests for bioactive nanoparticles: A common assay to detect reactive oxygen species as an example. *International Journal of Molecular Sciences*. 14:24320–24337 (2013).

23 Analysis of Cell Cycle Regulation and Apoptosis Using Annexin-V and Propidium Iodide (PI) Staining

Ravichandran Jayasuriya, Karan Naresh Amin, Sundhar Mohandas, and K.M. Ramkumar
Department of Biotechnology, School of Bioengineering, SRM Institute of Science and Technology, Kattankulathur, Tamil Nadu, India

23.1 INTRODUCTION

Every living cell undergoes a process of programmed cell death called apoptosis. It differs from necrosis, which is an accidental damage from morphological changes that appear in the membrane to the nuclear destruction and cytoplasmic shrinkage. In the cytoplasmic surface of the cell membrane in a living cell, a fatty substance or phospholipid is located and is called phosphatidylserine (PS). This PS then translocates from the inner side to the outside of the plasma membrane, which marks the cell for apoptosis by macrophages. Annexin is a Ca^{2+}-dependent human protein of about 35–36 kDa and has a high affinity towards PS. Thus, this anti-coagulant protein, which is labeled with FITC, or any other fluorescent dye binds to the PS and identifies the apoptotic cell population.

Propidium iodide (PI) is a red-fluorescent dye that binds to the DNA between the bases and hence is preferred to use in nuclear or chromosome staining. As it is non-permeable to live cells, they can also be used to differentiate live cells from the dead population. The and emission wavelength of PI are 535 nm and 617 nm, respectively. FITC excites at 495 nm and emits green color at 519 nm. FITC-Annexin-V staining at PS externalization means early apoptosis, whereas PI at the nuclear level means DNA damage (late apoptosis).

DOI: 10.1201/9781003397755-26

23.2 MATERIALS

- Cell culture facility
- six well cell culture plates
- Pipettes and micro tips (sterile)
- Tabletop centrifuge
- Annexin V/PI staining kit or dye
- Phosphate buffer saline (PBS)
- Hydrogen peroxide (H_2O_2)

23.3 REAGENT PREPARATION

1. 1X Annexin-binding buffer is prepared right before use for the respective number of experiments.
2. Prepare PI stock concentration 1 mg/mL, from which the working concentration of 100 µg/mL can be prepared in Annexin-binding buffer before use.

23.4 METHODOLOGY FOR ANALYSIS OF EARLY OR LATE APOPTOSIS

1. Plate approximately 0.2×10^6 cells per well in the appropriate medium with 10% FBS/ FCS on a sterile six-well culture plate.
2. Once the cells reach confluence, change the medium to no serum or low serum (0.5–1% FBS) for at least 2 hours prior to treatment.
3. The medium is aspirated, and fresh medium with 10–50 µM H_2O_2 is added, followed by an overnight incubation in a CO_2 incubator (H_2O_2 indices apoptosis).
4. The next day, the spent medium is aspirated, and the cells are washed with 1X PBS.
5. Trypsinize and harvest the cells by centrifugation in a sterile Eppendorf tube.
6. Spin down the cells, and resuspend in 1X Annexin V binding buffer/ solution. (*Note: 100 µL of binding buffer can be used for 1 million cells.*)
7. For 100 µL of binding solution used, add 5 µL of FITC conjugated Annexin-V and incubate for 1 hour in dark. This is followed by incubation with 1 µL of PI (working concentration).
8. Incubate the cells at room temperature for 15–20 minutes. (*Note: Store in a dark place.*)
9. Once the incubation is complete, directly mix another 400 µL to the samples tube and incubate on ice. (*Note: Do not resuspend in PBS or water; use only binding buffer/solution. The samples have to be maintained on ice until the next step.*)
10. Analyze the cells in a flow cytometer at the desired wavelength, as mentioned above.

11. Alternatively, the cells can be picturized using a fluorescent/confocal microscopy by adding 5–25 μL of FITC-Annexin-V and 1–2 μL of PI.

23.5 METHODOLOGY FOR ANALYSIS OF CELL CYCLE

1. Plate approximately 0.2×10^6 cells per well in the appropriate medium with 10% FBS/FCS on a sterile six-well culture plate.
2. Once the cells reach confluence, change the medium to no serum or low serum (0.5–1% FBS) for at least 2 hours prior to treatment.
3. Treat the cells with the desired compounds, and wash the cells with 1X PBS after treatment.
4. The cells are then trypsinised, fixed using 70% ethanol overnight, and resuspended in 100 μL PBS. An appropriate concentration of PI is added and incubated for 15–30 minutes at room temperature.
5. The tubes are then shifted to ice and maintained in dark until analysis.

23.6 INTERPRETATION OF EXPECTED RESULTS

In a flow cytometer, the results will be in the form of scatterplot, where more apoptotic cells can be observed at the right side with more fluorescent intensity. When using a double stain (FITC & PI), the X and Y axis can correspond to two different dyes, where the higher intensity means higher number of apoptotic cells. In a fluorescent microscope, the picture can be visualized with green and red colors corresponding to early and late apoptosis.

For analysis of cell cycle regulation, the peaks are looked at in the histogram. It provides the percentage of fractional DNA content. The cells that participate in G1, S, and G2/M phases of the cycle are also analyzed with the histogram. G1 phase means that the cells are ready for DNA replication. S phase denotes the replication of nuclear DNA, where the DNA content is doubled at the end of this phase. G2 denotes the completion of replication or the cells are ready for the onset of replication. The cells, if treated with a cytotoxic compound, tend to lock in G1 or sub-G1 phase, which infers the cells are not ready for replication [1–3].

REFERENCES

[1] Z. Darzynkiewicz, E. Bedner, and P. Smolewski. Flow cytometry in analysis of cell cycle and apoptosis. *Seminars in Hematology*. 38:179–193 (2001).
[2] M.G. Ormerod. Using flow cytometry to follow the apoptotic cascade. *Redox Report: Communications in Free Radical Research*. 6:275–287 (2001).
[3] C. Vignon, C. Debeissat, M.T. Georget, D. Bouscary, E. Gyan, P. Rosset, and O. Herault. Flow cytometric quantification of all phases of the cell cycle and apoptosis in a two-color fluorescence plot. *PLoS One*. 8:e68425 (2013).

24 Estimation of Chemokines and Cytokines Using Bio-Plex Assay

Ravichandran Jayasuriya, Karan Naresh Amin, Sundhar Mohandas, and K.M. Ramkumar
Department of Biotechnology, School of Bioengineering, SRM Institute of Science and Technology, Kattankulathur, Tamil Nadu, India

24.1 INTRODUCTION

The complexity of intracellular and extracellular signaling depends on the release of chemokines or cytokines. The release of these immune mediators offers a comprehensive understanding of the mechanism of disease progression and developing a therapeutic aid. The quantification of multiple cytokines from same-cell supernatant is challenging. This technique employs the quantification of a network of cytokines from same-sample or cell supernatant. This advancement is more useful when working with primary cultures or tissue samples, when the sample quantity is limited.

Multiplex is an ELISA-based approach where sample volume is not a concern. Traditional methods of ELISA allow the estimation of a single cytokine with a single sample where multiplex has changed them. The assays can be performed from a variety of sample types, including cerebrospinal fluid, saliva, and sputum other than cells and tissues. This advanced ELISA platform helps in the development of various drugs and vaccines.

24.2 MATERIALS

(As this is a kit-based method, the components and methodology differ with different manufacturers.)

The kit may contain the following:

- Wash buffer, assay buffer, detection antibody, beads, streptavidin conjugated (usually PE)

DOI: 10.1201/9781003397755-27

Apart from the kit contents, the following are required:

- Standards (BSA in the case of proteins)
- Shaker cum incubator
- Bio-Plex instrument
- Bio-Plex wash station

Before the start of the experiment, equilibrate or calibrate the following:

- Bio-Plex instrument (check for the validation of the instrument, ensure sufficient sheath fluid)
- Bio-Plex wash station (prepare the wash solution and fill the storage tank)

24.3 METHODOLOGY

1. Treat the cell lines or animals.
2. Collect and harvest the samples.
3. Wash the samples (if necessary).
4. In the case of animal or human tissue, homogenize using a tissue homogenizer in total cell lysis buffer at 4°C or on ice (*Note: Lysis buffer may contain protease inhibitor. If not, it is best to use one. Organalle specific protein if needed has to be seperated before using appropriate kit/lysis reagents*).
5. Once the lysis reagent is added, immediately transfer the contents to ice and mix using a vortex.
6. In case of hard tissues, an appropriate volume of zirconia/silica beads are added and homogenised using a bead-based homogeniser (Note: *The amptitude can be preferably set to 6.0 m/sec for 40 seconds. This varies with complexity of sample*).
7. In case of cultured cells, lysis by repeated freeze-thaw or using vortex would be sufficient (Note: *A homogenizer can be used with low settings recommended to lyse cultured cells*).
8. The homogenate is then centrifuged at 12,000 rpm (4°C).
9. To evaluate the release of cytokines/chemokines in delete to the culture medium, cell supernatant can be used directly or can be precepitated using acetone and resuspended using desired volume of sample diluent provided.
10. For quality control, QC buffer (provided in the kit) can be used with no further dilution (unless provided in the kit manual).
11. The supernatant of the homogenate is added to the detection antibody. (*Note: The antibody may be supplied concentrated. Dilute before use with the antibody diluent or use the appropriate volume as mentioned in the kit manual.*)
12. Seal the plate with adhesive film. (*Note: This step avoids the spillage of solution on incubation.*)

13. Incubate the plate at room temperature for 10–15 minutes (may vary depending upon the kit). Incubation may be preferred with gentle shaking not more than 100 rpm.

14. Remove the sealant and wash using wash buffer (three times) in the wash station (*Note: Manual aspiration may be avoided to minimize the risk of losing the bound antibodies.*)

15. Add the appropriate volume of streptavidin-conjugated antibody to the wells. (*Note: The antibody may be concentrated. Dilute the antibody using Bio-Plex assay buffer or any equivalents, if mentioned.*)

16. Seal the plate with a fresh adhesive film and cover with aluminum foil. (*Note: This step avoids the spillage of the sample and photobleaching of conjugated fluorophore.*)

17. Incubate the plate at room temperature for 10–15 minutes with gentle shaking not more than 100 rpm.

18. After incubation, remove the sealant and wash the plate with wash buffer (3 times) in the wash station.

19. Add the appropriate volume (usually 100–150 µl) of assay buffer, seal with fresh adhesive film, and incubate for 15 minutes with gentle shaking.

20. Remove the plate. The plate is now ready to read (using the appropriate settings provided with the kit or the default settings in the instrument's manual).

Interpretation of expected results:

The results are displayed in the form of values. Export the file and use it for analysis. The treatment group may be compared to control or equivalent to see the fold difference of any cytokine/chemokine [1–4].

REFERENCES

[1] N.J. Hannan, K. Bambang, T.J. Kaitu'u-Lino, J.C. Konje, and S. Tong. A bioplex analysis of cytokines and chemokines in first trimester maternal plasma to screen for predictors of miscarriage. *PLoS One*. 9:e93320 (2014).

[2] B. Houser. Bio-Rad's Bio-Plex(R) suspension array system, xMAP technology overview. *Archives of Physiology and Biochemistry*. 118:192–196 (2012).

[3] E. Karsten, E. Breen, and B.R. Herbert. Red blood cells are dynamic reservoirs of cytokines. *Scientific Reports*. 8:3101 (2018).

[4] M. Manglani, R. Rua, A. Hendricksen, D. Braunschweig, Q. Gao, W. Tan, B. Houser, D.B. McGavern, and K. Oh. Method to quantify cytokines and chemokines in mouse brain tissue using Bio-Plex multiplex immunoassays. *Methods*. 158:22–26 (2019).

Section 4

Cell Transfection and Single-Cell Analysis

25 Reporter Gene Construction and Assay

Karan Naresh Amin and Ravichandran Jayasuriya
Department of Biotechnology, School of Bioengineering,
SRM Institute of Science and Technology, Kattankulathur,
Tamil Nadu, India

R. Senthilkumar
Department of Biotechnology, School of Applied Sciences,
REVA University, Bangalore, Karnataka, India

K.M. Ramkumar
Department of Biotechnology, School of Bioengineering,
SRM Institute of Science and Technology, Kattankulathur,
Tamil Nadu, India

25.1 REPORTER ASSAY

Reporter assays are techniques that use reporter genes to quantify various biological activities. Reporter assays are a crucial range of methodology to study target gene expression and regulation. These strategies are widely used in the field of molecular biology, biochemistry, and biomedical and pharmacological research.

A reporter gene consists of a promoter linked to an exon sequence that transcripts the gene of interest. The promoter is responsible for regulating the expression of the gene as necessary. The reporter assay exploits this ability of the promoter region of the gene of interest by linking the promoter with a readily traceable and measurable reporter gene. Usually, the cells are transfected with the promoter-linked reporter gene into the cells, which are then introduced to various treatments or conditions whose effect on the gene expression/regulation can then be measured by changes in light emissions. The most common reporter genes include luciferases, fluorescent proteins, alkaline phosphatase, β-Galactosidase, β-Glucuronidase, and blue-white screening.

Luciferase, in particular firefly (*Photinus pyralis*) luciferase, is the most commonly used bioluminescent reporter assay as it is highly sensitive with a wide range, its light output can be readily measured, it is quick to perform, and it is cost-effective. The reaction takes place in two steps. First, the D-luciferin is activated in the presence of ATP to yield luciferyl adenylate and pyrophosphate, which is followed by the oxidation of luciferyl adenylate, which gives oxyluciferin in its higher energy state. The return of oxyluciferin from the high to low energy state is coupled with the slow release of green to yellow light (550–570 nm), which is detected with

DOI: 10.1201/9781003397755-29

the help of a luminometer. Other luciferase sources include various beetles, Renilla (sea pansy), and a modified deep sea shrimp.

Reporter assay techniques are typically performed to study alterations in promoter activity, but they can also be used to explore post-transcriptional activity, function of proteins, signal transduction, and microRNA regulation. Reporter assays can also be used to investigate and measure real-time changes in living organisms. Further, reporter assays can be used to assess apoptosis, assess metabolism, and analyze the effect of UTRs on mRNA stability, localization of protein, and translation efficiency.

25.2 DESIGNING A REPORTER GENE

A classic reporter gene should be designed in such a way that it should not be endogenously expressed in the cells of interest. It should be responsive to assay with high sensitivity, easy to quantify, quick and easy to perform, precise, reproducible, and safe. Reporter assays can be performed with one or even two reporter genes. The second reporter gene is usually used as control and co-transfected with the primary reporter gene for normalization of the results and can also be used for determining the transfection efficacy as well. Both the control and the primary reporter genes are linked to different promoter regions to avoid any signal interference between them. There are a few guidelines that can be followed to design and optimize the reporter gene sequence that is most suitable for the experiment.

- You can incorporate the proximal promoter region (250 bp to 1 kbp upstream of the promoter), the particular fragment of the promoter region, or even a single response element.
- You might also insert the transcription start site as well if it is unavailable on the chosen vector.
- Insert the 3' UTR region to examine post-transcriptional miRNA regulation.
- You can also include the 5' UTR region to study its effect on the target promoter activity.
- To study RNA splicing or regulation of the intron, you can include the intron along with part of its respective exon sequences.

25.2.1 FIRST STEP: LITERATURE REVIEW

The initial step is to research the gene of interest's sequence to be integrated into the reporter gene. The National Center for Biotechnology Information (NCBI) is an ideal place to search information about the gene of interest. There is also the possibility of finding a suitable vector that has already shown successful results for other researchers.

For a new gene sequence, a well-tested method is to include 1–3 kbp of your gene of interest and later try to reduce the size by incorporating different sections of the gene of interest to scale down to the key gene segment. Multiple copies of the known segment can be added to intensify the signal.

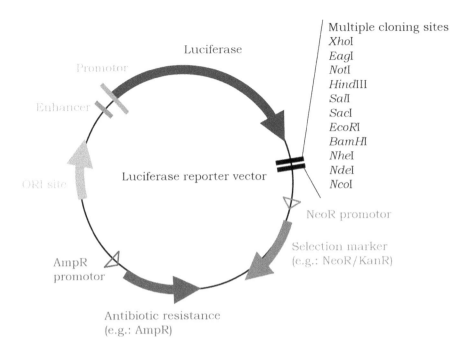

Multiple cloning sites
*Xho*I
*Eag*I
*Not*I
*Hind*III
*Sal*I
*Sac*I
*Eco*RI
*Bam*HI
*Nhe*I
*Nde*I
*Nco*I

FIGURE 25.1   Luciferase plasmid vector with ampicillin (bacterial selection) and neomycin (mammalian selection) selectable markers.

25.2.2   Second Step: Choose the Appropriate Vector

Once the gene sequence to be incorporated is finalized, the next step is to decide on the transfection vector. Usually a promoterless or a minimum promoter vector without inserts is used to reduce the level of background signaling, which will depend on the type of cells used and the transfection conditions. A promoterless vector is advisable to be used in case the gene of interest is a promoter sequence, whereas a minimum promoter vector is preferred if the gene of interest is anticipated to only amplify the transcription efficiency (Figure 25.1).

25.2.3   Third Step: Secondary Vector Attributes

Following the selection of the vector, you can then choose suitable vector modifications. For the development of stably transfected cells, you can use vectors with an antibiotic-resistance gene incorporated into it. Later, treating the cells with the respective antibiotic will eliminate the untransfected cells. Puromycin, hygromycin, neomycin, and other such antibiotics can be used based on the experiment and the kind of cells used.

In the case of studying gene regulation, it is better to alter the vector to possess a short half-life, as it will prevent signal escalation along with the gradual protein accumulation. The short half-life closely links protein expression with its corresponding transcriptional activity of the gene of interest. This method also helps increase the signal-to-background ratio, lower the assay time, and reduce any non-essential effects due to the extended incubation of cells with the screening compounds.

Also, there are commercially available vectors for commonly used target genes, which can be used provided they are feasible for your experiment.

26 Vector Transfection and Validation

Karan Naresh Amin and Ravichandran Jayasuriya
Department of Biotechnology, School of Bioengineering,
SRM Institute of Science and Technology, Kattankulathur,
Tamil Nadu, India

R. Senthilkumar
Department of Biotechnology, School of Applied Sciences,
REVA University, Bangalore, Karnataka, India

K.M. Ramkumar
Department of Biotechnology, School of Bioengineering,
SRM Institute of Science and Technology, Kattankulathur,
Tamil Nadu, India

26.1 FIRST STEP: BACTERIAL CELL TRANSFORMATION

The first step is to transform bacterial cells such as *E. coli* with plasmid DNA vectors incorporated with selectable markers such as antibiotic resistance. The most common technique is the heat shock method.

26.1.1 MATERIALS

- Microcentrifuge tubes
- Competent bacterial cells
- SOC medium (without antibiotic)
- LB agar plates (with appropriate antibiotic)
- Incubator with shaker
- Micropipettes

26.1.2 METHODOLOGY

1. Incubate the concoction of competent bacterial cells and DNA on ice for a short amount of time, then keep at 42°C for 45–60 seconds, and then again incubate on ice.
2. Following this, grow the transformed cells in SOC medium (without antibiotics) at 37°C with constant agitation.

DOI: 10.1201/9781003397755-30

3. Plate the cells on the LB agar plate containing an appropriate antibiotic and incubate at 37°C overnight. Only the transformed cells will be able to grow and form colonies.

4. Isolate the transformed cells and grow them until the desired concentration of transformed cells is obtained.

26.2   SECOND STEP: ISOLATION OF PLASMID VECTOR

Isolation of plasmid DNA from transformed bacterial cells using alkaline lysis is a well-established technique.

26.2.1   MATERIALS

- Microcentrifuge Eppendorfs
- Resuspension buffer (50 mM Tris-HCl, pH – 8, 10 mM EDTA, 100 μg/ml RNase A)
- Lysis buffer (0.2 N NaOH, 1% SDS)
- Neutralization buffer (3/5 M Potassium acetate, pH – 6)
- Spin column
- Isopropanol
- Elution buffer (water or TE buffer – 10 mM Tris, pH – 8, 1 mM EDTA)
- 75% Ethanol
- Micropipettes
- Vortex

26.2.2   METHODOLOGY

1. Take 1.5–2 ml of bacterial culture in a microcentrifuge tube and pellet the cells by centrifuging at 4°C for 2 minutes at 10,000 rpm.

2. Remove the supernatant, and to the pellet add 200 μl of resuspension buffer and vortex until the pellet is resuspended completely.

3. To the above suspension, add 250 μl of lysis buffer, and invert mix the suspension thoroughly until you have a clear viscous solution.

4. Add 350 μl of resuspension buffer and again invert mix thoroughly until you can see precipitate formation and centrifuge at 4°C for 10 minutes at 10,000 rpm.

5. Separate the supernatant and add 700 μl of chilled isopropanol. Incubate the tube at −20°C for 20 minutes.

6. Add an appropriate amount of the above solution into a spin column, centrifuge the column at 4°C for 1 minute at 8000 rpm, and discard the flow through. (Repeat this step until the entire volume of the above solution is used up.)

7. Then, add 400 μl of 75% ethanol, centrifuge the spin column at 4°C for 1 minute at 8000 rpm, and discard the flow through. (Perform this step twice.)

8. Centrifuge just the spin column at 4°C for 2 minutes at 10,000 rpm to flush out excess ethanol.

9. Further, place the spin column onto a new tube and add 30–50 µl elution buffer, and centrifuge the spin column at 4°C for 2 minutes at 10,000 rpm.
10. Store the eluate at −20°C for long-term storage.

26.3 THIRD STEP: CELL CULTURING AND SEEDING

Once sufficient concentration of the plasmid DNA vector is available, the next step is transfection of the vector into appropriate cell cultures. Prior to transfection, the relevant cells need to be cultured and maintained in aseptic conditions. (Refer to the cell culture section.)

26.3.1 METHODOLOGY

- The day of performing transfection, seed approximately $0.2 \times 10 \times 10^6$ (80–90% confluence) of relevant cells in a 24-well plate in 0.5 ml of appropriate serum-containing growth medium.
- Incubate the cells at 37°C with 5% CO_2 for 4 hours. (Incubation time will vary based on the type of cells used.)

26.3.2 FOURTH STEP: TRANSFECTION OF PLASMID VECTOR

Transfection of plasmid vector into appropriate cells can be carried out using either lipid or protein-based reagent methods.

26.3.3 MATERIALS AND EQUIPMENT

- Lipid or protein-based transfection reagent
- Cell culture medium (both with and without serum)
- Plasmid DNA vector
- Microcentrifuge tubes
- Micropipettes
- Vortexer
- CO_2 Incubator

26.4 METHODOLOGY FOR LIPID-BASED TRANSFECTION REAGENT

1. Warm up the cell culture medium (both with and without serum) to 37°C and the transfection reagent to room temperature in the dark.
2. In a microcentrifuge tube, add 50 µl of serum-free cell culture medium and 0.5 µg of plasmid DNA vector. Also, add 2 µl of lipid-based transfection reagent and vortex the mixture for 5–10 seconds.
3. Let this mixture sit in the dark for 30 minutes at room temperature.
4. Remove the old medium from the wells with seeded cells and add 250 µl of fresh cell culture medium supplemented with 10% serum to each well.

5. Following the 30-minute incubation, add 250 μl of fresh medium supplemented with 10% serum into the tube and mix thoroughly.
6. From the above mixture, add 300 μl to each well and incubate the plate for 3–4 hours at 37°C with 5% CO_2 incubator.
7. Following the incubation, remove the transfection medium and add 500 μl of fresh complete culture medium and culture the cells in a 37°C, 5% CO_2 incubator for 24–48 hours.

26.5 METHODOLOGY FOR PROTEIN-BASED TRANSFECTION REAGENT

1. Warm up the cell culture medium (both with and without serum) to 37°C.
2. In a tube, add 18.75 μl of serum-free cell culture medium and 0.5 μg of plasmid DNA vector. Also, add 0.75 μl of protein-based transfection reagent and vortex the mixture for 5–10 seconds.
3. Let this mixture sit for 10 minutes at room temperature.
4. Remove the old medium from the wells with seeded cells, and add 250 μl of fresh cell culture medium supplemented with 10% serum to each well.
5. Following the 10-minute incubation, add 250 μl of fresh cell culture medium supplemented with 10% serum into the tube and pipette mix thoroughly.
6. From the above mixture, add 270 μl to each well and incubate the plate for 3–4 hours at 37°C on a 5% CO2 humidified incubator
7. Following the incubation, remove the transfection medium and add 500 μl of fresh complete culture medium and culture the cells in a 37°C, 5% CO_2 incubator for 24–48 hours.

26.6 FIFTH STEP: STIMULATION OF CELLS AND LUCIFERASE REPORTER ASSAY

Once we obtain the relevant cells transfected with the desired reporter gene vector, the next step is to stimulate the cells with different treatment conditions and perform the reporter assay. For this section, we will focus on the luciferase reporter assay.

26.6.1 MATERIALS AND EQUIPMENT

- Cell culture medium
- Relevant cell stimulant
- Lysis buffer
- 2x Luciferase buffer (40 mM Tris-phosphate, 2.14 mM $MgCl_2$, 5.4mM $MgSO_4$, 0.2 mM EDTA, 66.6 mM DTT, pH – 7.8)
- Luciferase assay reagent (For 1 ml)
  - 500 μl 2× luciferase buffer
  - 47 μl luciferin 10^{-2} M
  - 53 μl ATP 10^{-2} M
  - 27 μl Coenzyme A (lithium salt) 10^{-2} M

- 430 μl distilled water
- Microcentrifuge tubes
- Micropipettes
- Vortexer
- CO_2 Incubator

26.7 METHODOLOGY

- Discard the old media and add fresh cell culture medium in each well.
- Add the appropriate concentration of cell stimulant relevant to your experiment, and keep the plate in the 37°C, 5% CO_2 incubator for the appropriate amount of time.
- Following the respective treatment, remove the spent cell culture medium and wash with chilled 1x PBS, after which add 200 μl of 1x Lysis buffer.
- Scrape the wells thoroughly and collect the lysate in a microcentrifuge tube. Centrifuge the tubes at 20,000 × g for 10 minutes at 4°C.
- Separate 50 μl of the clear supernatant in a clear bottom 96-well plate, and add 3 μl of the luciferase assay reagent (light sensitive) or follow the manufacturer's instructions for commercially available luciferase.
- Capture and interpret the luciferase signal through the visible light spectrum using a luminometer [1–3].

REFERENCES

[1] Neefjes, M., Housmans, B.A.C., van den Akker, G.G.H. *et al.* Reporter gene comparison demonstrates interference of complex body fluids with secreted luciferase activity. *Sci Rep* **11**, 1359 (2021). 10.1038/s41598-020-80451-6

[2] Liu, A.M.F., New, D.C., Lo, R.K.H., Wong, Y.H. (2009) Reporter Gene Assays. In: Clemons P., Tolliday N., Wagner B. (eds.) *Cell-Based Assays for High-Throughput Screening. Methods in Molecular Biology (Methods and Protocols)*, vol. 486. Humana Press, Totowa, NJ. 10.1007/978-1-60327-545-3_8

[3] Schenborn, E., Groskreutz, D. Reporter gene vectors and assays. *Mol Biotechnol* **13**, 29–44 (1999). 10.1385/MB:13:1:29

27 Single-Cell Analysis

Karan Naresh Amin and Ravichandran Jayasuriya
Department of Biotechnology, School of Bioengineering,
SRM Institute of Science and Technology, Kattankulathur,
Tamil Nadu, India

R. Senthilkumar
Department of Biotechnology, School of Applied Sciences,
REVA University, Bangalore, Karnataka, India

Md Enamul Hoque
Department of Biomedical Engineering, Military Institute of
Science and Technology, Dhaka, Bangladesh

K.M. Ramkumar
Department of Biotechnology, School of Bioengineering,
SRM Institute of Science and Technology, Kattankulathur,
Tamil Nadu, India

27.1 INTRODUCTION

Single-cell analyses are the latest techniques to explore cellular heterogeneity between individual cells in different biological systems. In most of the cell studies population-averaged results are measured, but at times measurements from some of the relevant subpopulation of cells present in the system are overwhelmed, especially in conditions where the subpopulation regulates the behavior of the entire population [1]. Information associated with cell-to-cell differences in RNA transcripts and protein expression can be vital in explaining many unexplained phenomena throughout various branches of biology. Studying individual cell phenotypes is useful in understanding precise gene expression measurements, protein analysis, and signaling response growth dynamics, and it also provides more detailed information for therapeutic decision-making in precision medicine. However, to date, single-cell studies have been limited by the cost and throughput required to examine large number of cells and the difficulties associated with analyzing small amounts of starting material. There are many approaches to single-cell analysis, including single-cell genomics, single-cell transcriptomics, and single-cell proteomics. Some of the known techniques to isolate and perform single-cell analysis include fluorescence-activated cell sorting (FACS), magnetic-activated cell sorting, laser capture microdissection, manual cell picking/micromanipulation, microfluidics, linker-adapter PCR, the multiple

DOI: 10.1201/9781003397755-31

displacement amplification, single-cell RNA sequencing and microarrays, mass spectrometry, etc.

FACS is a type of flow cytometry with the facility of sorting cells. It is one of the most widely used techniques for characterizing and defining different cell types in a heterogeneous cell population based on size, granularity, and fluorescence. FACS has the capacity to perform simultaneous quantitative and qualitative multi-parametric analyses of single cells. Briefly, the target cells are labeled with fluo-rescent probes that recognize specific surface markers on target cells. As the cells run through the cytometry, each cell is exposed to the respective laser, which allows the fluorescence detectors to identify cells based on the selected characteristics. The cells of interest can then be sorted in a respective collection tube using an elec-trostatic deflection system. Single-cell proteomics using FACS can be used to identify protein concentration, location, post-translational modifications, or inter-actions with other proteins [2].

27.2 FACS-BASED SINGLE-CELL PROTEIN ANALYSIS

27.2.1 Materials

- Cell line of interest
- Cell culture medium
- Stimuli of interest
- 16% Paraformaldehyde (PFA)
- 100% Methanol
- Flow cytometry staining buffer
- Appropriate fluorescent labeled-specific antibodies
- 5-ml polystyrene FACS tubes (BD Falcon)
- 37°C, 5% CO_2 incubator
- Centrifuge
- Flow cytometer with respective laser lines and sorter

27.2.2 Methodology

The protocol followed here is modified based on [1–3]

Preparation of Cell Culture Samples

1. Grow the cell line of interest in an appropriate cell culture medium sup-plemented with 10% FBS and add stimuli/treatment of interest to your cells. Incubate the cells at 37°C, 5% CO_2 incubator for an appropriate amount of time.
2. Detach the cells accordingly and centrifuge the cell suspension at $375 \times g$ for 10 minutes, discard the supernatant and resuspend in the cell culture medium supplemented with 10% FBS.
3. Count the number of cells using a hemocytometer under a microscope and adjust the cell density to 10^6–10^7 cells/ml.
4. Distribute 1 ml to each 5-ml FACS tube.

5. Fix the cells by adding 100 µl of fresh 16% PFA per ml of medium, vortex mix for few seconds and incubate for 10 min at RT. Centrifuge for 5 minutes at 375 × g, 4°C, and decant the supernatant.
6. Add 1 ml of chilled methanol to each tube, and vortex thoroughly.
7. Incubate the tubes for 20 minutes at 4°C.

Staining cells with appropriate fluorescent labeled-specific antibodies:

1. Add 3 ml of Flow cytometry staining buffer to the tubes for washing and centrifuge at 375 × g for 5 minutes at 4°C, and decant the supernatant. (Repeat this step again.)
2. Resuspend the cells in Flow cytometry staining buffer at ~1 × 10^6 cells/60 µl and transfer 60 µl to fresh tube.
3. Add 20 µl appropriate concentration of fluorescent labeled-specific antibodies and incubate in a dark place for 30 minutes in RT.
4. Pellet the cells, wash twice, and resuspend in Flow cytometry staining buffer as shown earlier.
5. Analyze using a flow cytometer.

27.2.3 FACS SAMPLE ANALYSIS

1. First acquire events with side-scatter area linear scale (SSC-A-lin) v/s linear forward-scatter area linear scale (FSC-A-lin) axes the unstained or unlabeled sample.
2. Adjust the acquisition speed to 100–200 events per second, which should show a majority of the cell population in the SSC v/s FSC plot, and with the help of this plot, center the acquired events using the voltage and gain settings to create the first gate.
3. Next, isolate cell doublets by plotting FSC-width-lin v/s FSC-A-lin and set gate R2 around single cells. If there is any fluorescent probe used for DNA content, isolate cell doublets by plotting a linear 405 nm area (405-A-lin) versus a linear 405 nm height (405-H-lin), and set gate R2 around single cells and change the 405-nm laser voltage to center the population between 50,000 and 100,000.
4. Plot the detected fluorescence in histogram mode: counts versus logarithmic respective wavelength area, based on the type of fluorescence probe used, and change the corresponding voltage to center the background peak at around 10^2.
5. Save the respective settings on the flow cytometer.
6. First run the control or unstimulated cell samples acquiring at least 5000–10,000 events.
7. In the histogram plot, establish an arbitrary gate to define positive cells.
8. Similarly, run the stimulated cell samples.
9. After the run, export the data and perform post-acquisition analysis using appropriate software such as FlowJo.

REFERENCES

[1] Perez, O., and Nolan, G. Simultaneous measurement of multiple active kinase states using polychromatic flowcytometry. *NatBiotechnol.* **20**, 155–162 (2002). 10.1038/nbt0202-155

[2] Hu, P., Zhang, W., Xin, H., and Deng, G. Single cell isolation and analysis. *Front. Cell Dev. Biol.* **4**, 116 (2016). 10.3389/fcell.2016.00116

[3] Forment, J., and Jackson, S. A flow cytometry-based method to simplify the analysis and quantification of protein association to chromatin in mammalian cells. *Nat. Protoc.* **10**, 1297–1307 (2015). 10.1038/nprot.2015.066

Index

For Product Safety Concerns and Information please contact our EU
representative GPSR@taylorandfrancis.com
Taylor & Francis Verlag GmbH, Kaufingerstraße 24, 80331 München, Germany

www.ingramcontent.com/pod-product-compliance
Ingram Content Group UK Ltd.
Pitfield, Milton Keynes, MK11 3LW, UK
UKHW021123180425
457613UK00005B/196